Nuclear Chemistry

Maheshwar Sharon · Madhuri Sharon

Nuclear Chemistry

Second Edition

Ane Books
Pvt. Ltd.

Maheshwar Sharon
Walchand Centre for Research in
Nanotechnology & Bionanotechnology
Walchand College of Arts and Science
Solapur, Maharashtra, India

Madhuri Sharon
Walchand Centre for Research in
Nanotechnology & Bionanotechnology
Walchand College of Arts and Science
Solapur, Maharashtra, India

ISBN 978-3-030-62020-2   ISBN 978-3-030-62018-9   (eBook)
https://doi.org/10.1007/978-3-030-62018-9

Jointly published with ANE Books Pvt. Ltd.
In addition to this printed edition, there is a local printed edition of this work available via Ane Books in
South Asia (India, Pakistan, Sri Lanka, Bangladesh, Nepal and Bhutan) and Africa (all countries in the
African subcontinent).
ISBN of the Co-Publisher's edition: 9789386761552

This Springer imprint is published by the registered company Springer Nature Switzerland AG
The registered company address is: Gewerbestrasse 11, 6330 Cham, Switzerland

— Acharya Kanad
— (Founder of Atomic Theory)

*As the founder of "Vaisheshik Darshan"—
one of six principal philosophies of India—
Acharya Kanad was a genius in philosophy.
He is believed to have been born in Prabhas
Kshetra near Dwarika in Gujarat. He was the
pioneer expounder of realism, law of
causation and the atomic theory. He has
classified all the objects of creation into nine
elements, namely: earth, water, light, wind,
ether, time, space, mind and soul.*

*He says, "Every object of creation is made of
atoms which in turn connect with each other
to form molecules". His statement ushered in
the Atomic Theory for the first time ever in the
world, nearly 2500 years before John Dalton.
Kanad has also described the dimension and
motion of atoms and their chemical reactions
with each other.*

*The eminent historian, T. N. Colebrook, has
said, "Compared to the scientists of Europe,
Kanad and other Indian scientists were the
global masters of this field".*

Foreword

I have read your manuscript with great interest. I am sure that the book will be well received as a clear and concise treatment of radiation detection and measurement. The book is also attributed in large measure to teach undergraduate, graduate and other educational courses.

Byung-Hun Lee
Professor of Nuclear Engineering
Department of Nuclear Engineering
College of Engineering
Hanyang University
Seoul, Korea

Preface

We are glad to learn that this book has been appreciated by readers so much that its 2nd edition is published. In this 2nd edition, we have added some new topics considering the syllabus of the M.Sc. course dealing with Nuclear Chemistry like the Shell model, fission/fusion reaction, natural radioactive equilibrium series, nuclear reactions carried out by various types of accelerators, etc. It is hoped that this 2nd edition will be appreciated by readers and graduate and postgraduate students.

It has been our experience that students of different disciplines desirous of using the tracer technique face the problem of selecting an isotope most suitable for their work. Having selected the isotope, the next problem they face is in selecting the type of counter, type of source sample, duration for which the counting must be made, and selecting the radiation emitted by the isotope for its efficient detection. This book is an effort toward guiding the readers on these aspects. Though there are many books available on this topic, they are either very advanced or very elementary. Students have, therefore, been requesting the author for many years to write a book of this nature. This book is, thus, an outcome of the lectures given to Engineering and Science graduate and postgraduate students who have had very little exposure to the field of radioactivity. It has been tried to explain topics from the experience the author gained while teaching this subject in England and India. The author has also given a few results obtained by him while developing some experiments for students studying this course.

It has been the experience that students often resort to memorizing the decay process without actually understanding the logic of various decay processes. Similarly, it is observed that students invariably select a counting procedure without understanding the reasons behind its selection. A novel concept is, therefore, developed to explain not only the decay processes but also the selection of counting procedures for the detection and measurement of radioactivity. It is hoped that students would get an exposure to select the counting procedures rather than simply getting bookish knowledge of the subject. This book, thus, concentrates on the techniques concerned with the detection and measurement of radioactivity. However, in order to appreciate the subject, an introduction to properties of radioactivity, e.g., law of decay of radioactivity, type of decay, and interaction of radiation including particulate radiation with matter are dealt with appropriately. This book builds up its foundation from the nature of the interaction of radiation with matter, which brings out the differentiations among ionization counter, scintillation counter, and solid-state detector. Based on decay properties like nature of radiation, its energy, and abundances (which are given in the decay scheme), an attempt is made to select the type of counter for measuring radioactivity. Similarly, the selection of the form in which a sample should be prepared for counting is also discussed. The statistics of counting is enumerated, which can assist to decide some of the routine-type corrections like subtraction and division of standard deviation calculation. Finally, an effort is made to make the reader aware that there is a difference between an ordinary chemical laboratory and a radiochemical one. There is a need to understand that in designing a radiochemical laboratory, various considerations are needed which normally are not essential for an ordinary chemical laboratory. Accordingly, these aspects of design of a radioactive laboratory have been discussed. At the end, various possible questions related to counting are mentioned so that the reader after solving them would become more familiar with this field.

It is hoped that the 2nd edition of the book would suffice the need of readers. However, it would be the pleasure of the authors to receive valuable comments and suggestions from readers so that the next edition could be further improvised.

Solapur, India   Maheshwar Sharon
                                                                                sharon@iitb.ac.in
                                                                                Madhuri Sharon
                                                                          sharonmadhuri@gmail.com

Contents

About the Authors

Late Prof. Maheshwar Sharon obtained Postgraduate Advance Diploma in Nuclear Power from Strachclyde University, Glasgow (1962); Postgraduate Diploma (Radiochemistry), Leicester Polytechnic (1964) and Ph.D. from Leicester University (1967). At Manchester University and Bolton Institute of Technology, he was a Postdoctoral Fellow where he studied (n, y) reactions using University Research Thermal Reactor at Risely, UK. After a brief spell at Himachal and Pune Universities, respectively, he joined as a Professor in Chemistry at IIT, Bombay, in 1978 and retired in 2003. Thereafter he joined Birla College, Kalyan, as an Adjunct Professor and started a Nanotechnology Research Center. He was awarded Professor Emeritus (CSIR) 2003–2005. He was awarded Professor Emeritus by the UGC. He had been awarded the Man of the Year 1990, 1992, 1994 and 1996 by AB Inc, USA and was a Fellow of Royal Society of Chemistry, London (1965–2003). He had more than 150 research publications in National and International journals. He had many international research collaborations, namely with Japan, France and UK (with Prof. H. W. Kroto, FRS, Nobel Laureate). Unfortunately, he passed away in December 2020, while this book was in production.

Dr. Madhuri Sharon a Ph.D. from University of Leicester (1969), has more than four decades of teaching and research experience. She also worked as a Postdoctoral Fellow at Bolton Institute of Technology, UK, and joined Bolton Technical College, UK, as a teaching faculty. After returning to India, she joined as CSIR Pool Officer and worked at Sabour Agricultural University and Central Potato Research Institute. She joined as a Lecturer in Department of Botany, University of Pune, in 1973, where apart from teaching Masters students, she carried out research for 7 years. She has held senior positions in the industries like Gufic (as a Vice-President) and Reliance Industries (as Director).

Chapter 1
Nuclear Chemistry

1.1 Introduction

This chapter about Nuclear Chemistry encompasses a brief introduction to atoms, structure of a nucleolus, stability of elements, applications of the binding energy of nucleus such as fission and fusion reaction, etc.

1.2 Atom

The concept of the orbital electron configuration of an atom is presented in many textbooks of chemistry and need not be explained here in detail. Nevertheless, there are two main constituents of atoms: orbitals in which electrons are housed and a nucleus in which species like neutrons and protons are housed. Nuclear chemistry deals with the behavior of neutrons and protons, and sometimes with the behavior as well as the structure of electrons. Electrons are arranged outside the atomic nucleus in several orbitals. Each orbital, described by four quantum numbers, has definite energy and can accommodate two electrons with an antiparallel spin. The transition of an electron from one orbital to the other results in the emission or absorption of a definite amount of energy. The magnitude of this energy corresponds to the difference between the energy of the two orbitals concerned. This energy is emitted in the form of electromagnetic radiation.

Atoms combine together to form a compound, by the transfer of electrons from one atom to another followed by an electrostatic attraction (electrovalency) or sharing electrons (covalency). During chemical reactions, electrons present in the outermost orbitals are rearranged. Therefore, in the foregoing sections, our discussions are limited to the behavior of constituents of the nucleus, which mainly include neutrons and protons, their arrangements in the nucleus, and their impact on the stability of the nucleus.

M. Sharon and M. Sharon, *Nuclear Chemistry*,
https://doi.org/10.1007/978-3-030-62018-9_1

1.3 Nuclear Structure

In order to completely establish the constitution of an atomic nucleus, it is necessary to know the nature of the constituents, forces binding them together, and laws
which govern their behavior. The fundamental nuclear constituents are **protons** and
neutrons, and laws governing their interactions are those of quantum mechanics.
However, the precise nature of the nuclear forces is not yet fully understood. Much
of the known information are embodied in several nuclear models; each of these has
advantages, but none of them are able to explain all of the available experimental
data of a nucleus.

1.4 Shell Model

Among the various nuclear models, the shell model is more popular because it can
explain most of the nuclear behavior of the atom and has helped to synthesize new
isotopes, which were unknown to exist. The basis of this model follows almost similar
concepts as the arrangement of electrons in an atom. Therefore, it may be easier to
explain the shell model by comparing it with an atomic electron structure. Hence, it
would be perhaps better to briefly revive the basic knowledge of an atomic structure.
An atom consists of electrons and it revolves around the nucleus with different orbitals
depending upon the number of electrons present in the atom. Protons and neutrons
are present in the nucleus. The various orbitals designated for electrons are **s** which
can accommodate a maximum of 2 electrons, **p** which can house a maximum of
6 electrons, **d** which can accommodate a maximum of 10 electrons, and **f** orbital
accommodating 14 electrons as its maximum capacity. Atoms with total number of
electrons, i.e., 2 (Helium), 10 (Neon), 18 (Argon), 38 (Krypton), 54 (Xenon), and 86
(Radon) are the most inert and stable elements, and other elements containing less
or more number of electrons are usually less stable. These numbers are also referred
to as a **magic number**.

Like the electron arrangement in an atom, the shell model also assumes the revo-
lution of proton or neutron in some specific orbital such that not more than 2 nucleons
(i.e., either proton or neutron) would be present in one orbital. As nucleons are added
to the nucleus, they first occupy the lowest energy as allowable by the Pauli Exclu-
sion Principle. This means that each nucleon possesses a unique quantum number
which portrays its motion. When the orbital of the shell is completely filled, the
nucleus attains high stability. It is assumed that like electrons, nucleons possess spin
($\frac{1}{2}$) and they possess a quantum number like l and m. It is also assumed that within
the nucleus, separate orbitals exist for protons and neutrons.

Thus, the first two protons will be filled in the zero level (i.e., 0, 0, 0, $+\frac{1}{2}$ and
$-\frac{1}{2}$ = 2 protons). Likewise the next six protons will be filled in level one. In this
fashion, protons can be filled in all other possible levels. Considering the arrangement
of six shells, the number nucleon distributed would follow as shown here:

∗ level 0:   2 states ($l = 0$) = 2
∗ level 1:   6 states ($l = 1$) = 6
∗ level 2:   2 states ($l = 0$) + 10 states ($l = 2$) = 12
∗ level 3:   6 states ($l = 1$) + 14 states ($l = 3$) = 20
∗ level 4:   2 states ($l = 0$) + 10 states ($l = 2$) + 18 states ($l = 4$) = 30
∗ level 5:   6 states ($l = 1$) + 14 states ($l = 3$) + 22 states ($l = 5$) = 42

If nucleons are filled in this fashion for all possible shells, it is observed that some shells contain some specific number of nucleons as shown here:

⋆ **2**
⋆ **8** = 2 + 6
⋆ **20** = 2 + 6 + 12
⋆ **28** = 2 + 6 + 12 + 8
⋆ **50** = 2 + 6 + 12 + 8 + 22
⋆ **82** = 2 + 6 + 12 + 8 + 22 + 32
⋆ **126** = 2 + 6 + 12 + 8 + 22 + 32 + 44
⋆ **184** = 2 + 6 + 12 + 8 + 22 + 32 + 44 + 58

If protons and neutrons are filled in this fashion, it is observed that those nuclei which have fully occupied shells will have the number of nucleons as 2, 8, 20, 28, 50, 82, 126, and 184. It is observed that the nucleus containing these specific number of nucleons are highly stable compared to those which have one less or one more nucleon. This number is recognized as **Magic Number**.

The shell model thus predicts the possibility of new isotopes which have not yet been discovered like isotopes of magic numbers 184 and 126. It is observed that nuclei containing even numbers of protons and neutrons are more stable than those with odd numbers, which is due to the pairing effect. Moreover, nuclei which have both neutron and proton numbers equal to one of the magic numbers are called **doubly magic**, and are very stable. For example, Calcium ($^{40}Ca_{20}$, $^{48}Ca_{20}$) is a good example to explain doubly magic number nuclei.

The shell model, thus, has given a better understanding of why some isotopes are more stable than others and why a large number of isotopes are found with some specific nucleon number. The shell model has been able to explain some properties of a nucleus as enumerated here:

1. Large number of isotopes are found for nuclei of a magic number.
2. All naturally occurring radioactive series like Uranium 238 end with a stable isotope Pb 206, and all of them belong to the magic number of neutrons or protons.
3. Every isotope of nuclei has some specific values for the cross-section of neutron absorption. Those isotopes belonging to magic numbers have a lower cross-section than their surrounding isotopes.
4. Isotopes containing the last neutron of the neutron magic number nuclei have the maximum binding energy. If one more neutron is added to it then the binding energy of the isotope drops down sharply.
5. Nuclei of a magic number have nearly zero electric quadruple moments.

6. Excitation energy needed for nuclei belonging to the magic number is greater for its first excitation state.

The stability of isotopes can be better understood by a mathematical model. In general, an element "**X**" is represented as $^{A}\mathbf{X}_{Z}$, where A is the total mass of the element (i.e., the total number of protons and neutrons) and $\mathbf{Z}$ is the atomic number (i.e., the total number of protons).

Useful information can be obtained by studying the so-called binding energy which controls the stability of nuclei. A proton ($^{1}H_{1}$) has a unit positive charge and mass of one H atom (i.e., one proton and one electron), viz., 1.00782522 a.m.u., while a neutron ($^{1}n_{0}$) is an electrically neutral particle of mass slightly greater than that of the proton, viz., 1.00866544 a.m.u. Like charges repelling each other, the forces which bind protons and/or neutrons together in a nucleus must be other than an electrostatic force.

1.5   Binding Energy of Nucleus

It is observed that the mass of an atom is less than the combined masses of separate constituents of the atom. This difference in mass is called the "**mass defect**" and may be expressed in terms of "energy" by using the well-known Einstein's mass–energy relation, i.e.,

$$E = mc^2,$$

where "c" is the velocity of light and "m" the mass. This energy is responsible for binding nucleons together in the nucleus and is called the "**binding energy**".

For example, a He atom ($^{4}He_{2}$) with a mass of 4.00260361 a.m.u. can be considered to explain the binding energy. The separate constituents of He may be considered to be two H atoms (or two protons and two electrons) and two neutrons. The mass of these particles, sometimes called "**nucleons**", is expressed either in the unit of kg or in atomic mass unit (a.m.u.). One atomic mass unit (a.m.u.) $= 1.66054 \times 10^{-27}$ kg. This mass can be converted into energy by using Einstein's relation ($E = mc^2$). If mass is expressed in kg, and velocity of light in ms^{-1}, then energy (in the unit of mega electron volt, MeV) produced from one unit of a.m.u. can be obtained by multiplying mc^2 by a conversion factor of 62.41539×10^{-5}. Thus,

$$1 \text{ a.m.u.} = (1.66054 \times 10^{-27} \text{ kg}) \times (2.99792458 \times 10^{8} \text{ m/s or ms}^{-1})$$
$$\times 2 \times (62.41539 \times 10^{-5})$$
$$= 931.5 \text{ MeV}.$$

Hence, the binding energy of a He atom can be calculated as

Mass of one proton	$= 1.00782522$ a.m.u
Thus mass of two protons (2^1H)	$= 2.0156504$ a.m.u
Mass of one neutron ($^1n_0$)	$= 1.00866544$ a.m.u
Thus mass of two neutrons	$= 2.0173308$ a.m.u
Mass of one electron	$= 0.000548$ a.m.u
Mass of two electrons	$= 0.001096$ a.m.u
Mass of $2e^- + 2^1n_0 + 2^1$H$_1$	$= 4.0329812$ a.m.u
In other words, expected mass of He atom	$= 4.0329812$ a.m.u
Actual mass of He atom (^{4}He$_2$)	$= 4.00260361$ a.m.u
The loss in mass of He atom, i.e., mass defect	$= 0.0303776$ a.m.u

Therefore, the energy equivalent to mass defect in a He atom $= 931.5 \times 0.0303776 = 28.29$ MeV.

Thus, the binding energy of a He atom is 28.29 MeV; and this energy is required to breakup the atom into its constituents. The average binding energy per nucleon is 28.29 MeV/4 $= 7.07$ MeV. This means that in order to keep two protons and/or two neutrons in their stable form, an energy equivalent to 7.07 MeV per nucleon is involved. In other words, the higher the mass defect, the higher is the binding energy of the nucleus (or of nucleons). If this idea is further extended, then it can be concluded that when a radioactive (i.e., unstable nucleus, called parent nucleus) material is to decay to become a stable isotope (daughter nucleus), difference in the mass of the two (i.e., between the daughter and parent nuclei) should approximately be either equal to or greater than the multiple of binding energy per nucleon (i.e. $\approx$ multiple of 7.07 MeV). If the mass difference is less than this energy, then the product will not be stable nuclei. We can understand the advantages of this calculation better by considering the variation in the binding energy per nucleon for elements of various mass numbers.

A plot of average binding energy per nucleon against the mass number for naturally occurring nuclides is shown in Fig. 1.1. It will be noticed that four atoms, e.g., ^{4}He$_2$, ^{12}C$_6$, ^{16}O$_8$, and ^{20}Ne$_{10}$ do not lie on the curve, because their binding energies are greater than those expected from the smooth curve. In other words, these nuclei are more stable than other neighboring nuclei. Elements having mass numbers between 40 and 120 have the highest average binding energy per nucleon, of about 8.5 MeV and are most stable naturally occurring nuclides. This value decreases for higher mass numbers and finally comes to about 7.6 MeV for uranium. This suggests that uranium will have a tendency to break its nucleus into elements with binding energy greater than 7.6 MeV and the excess energy.

For example, 8.5 MeV $-$ 7.6 MeV $= 0.9$ MeV per nucleon is released in the form of electromagnetic radiation. It is this a diminution in binding energy, which is released during the fission of uranium. Similarly, thermonuclear energy (known as **fusion reaction**) is released by synthesizing higher mass number nuclides from lower ones, e.g., the fusion of H or He produces a similar reaction. In the fusion process, binding energy per nucleon is increased and excess energy is released in the form of electromagnetic radiation. The fusion products are normally stable isotopes (because their binding energies lie near maxima of the curve, i.e., about 8 MeV), whereas fission reaction products are mostly radioactive (because their binding energies lie much below the maxima of the curve). We shall now discuss briefly the Fission and Fusion processes.

1.6   Fission Reaction

It is realized from Fig. 1.1 that heavier atoms like $^{235}U_{92}$ with a binding energy of about 7.6 MeV per nucleon will have the tendency to break down easily to other mass which has higher binding energy than 7.6 MeV, like elements between 40 and 120 which have the average binding energy per nucleon of about 8.5 MeV. This means that excess energy, i.e., (8.5 MeV − 7.6 MeV) = 0.9 MeV will be realized per nucleon. It was observed that when $^{235}U_{92}$ absorbs one neutron to form $^{236}U_{92}$, it immediately gets decomposed to the following products:

$$^{236}U_{92} \rightarrow \, ^{141}Ba_{56} + \, ^{92}Kr_{36} + 3\,^{1}n_{0} \qquad (1.1)$$

It was observed that if this reaction is carried out in an enclosed system, a sequence of reactions continues facilitated by the 3 neutrons released per reaction. This was the first reaction which gave the idea of a spontaneously initiated reaction and was named as the **Fission** reaction. It was also observed that the efficiency of this reaction

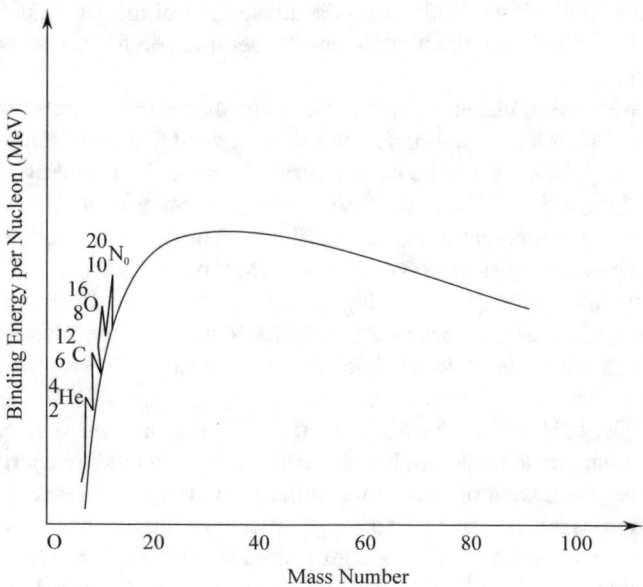

Fig. 1.1 Binding energy per nucleon for stable nuclides

improved if a neutron released from the fission is slowed down by suitably absorbing its energy by material like heavy water or graphite. These materials are also called **moderators**. They do not absorb the neutron but allow it to undergo several collisions within the material and what comes out is a slow neutron. Absorbing capacity, the so-called **cross-section** for the absorption of neutron, becomes more with a fresh uranium atom to produce the reaction −1. Energy released in this fission reaction is about 6.8×10^7 kJg^{-1} of ^{135}U$_{92}$. When fission with a heavy element takes place, many new elements are produced with a mass number ranging 72–161. Many of these elements are radioactive. The first destructive application of this fission reaction was unfortunately carried out by the US Army by dropping the so-called atom bomb on Hiroshima and Nagasaki in Japan. However, soon it was realized that if the concentration of neutrons interacting with uranium could be controlled then one can get energy for the benefit of mankind. This gave birth to a **nuclear reactor**. The most efficient fissionable material ^{235}U$_{92}$ in the form of uranium oxide is extracted from natural uranium (i.e., ^{238}U$_{92}$). Two kinds of reactors are used: deuterium (heavy water) reactor where heavy water acts as a coolant as well as a moderator, and gas-cooled reactors where graphite is used as a moderator. In order to maintain an equilibrium between the number of fusions taking place and the number of neutrons released, an efficient neutron absorbing material like boron is used to control the fission reaction. These combinations gave birth to the **thermal nuclear reactor** in which ^{235}U$_{92}$ is consumed to produce energy as well as many radioactive elements. These radioactive isotopes have many applications, which are discussed in forthcoming chapters.

Another type of reactor is also being developed known as the **breeder reactor**. It operates with high energy neutrons and produces more fissionable atoms than it can consume. Naturally occurring uranium contains mainly ^{238}U$_{92}$ and very small amount (approximately around 0.7%) of ^{235}U$_{92}$. In a breeder reactor, naturally occurring Uranium-238 is used which produces ^{239}Pu$_{94}$ by the following process:

$$^{1}n_0 + {}^{238}U_{92} \rightarrow {}^{239}U_{92} \rightarrow {}^{239}Np_{93} \rightarrow {}^{239}Pu_{93}$$

^{239}U$_{92}$ has a half-life of 24 min and decays with β-, and ^{239}Np$_{93}$ decays with a half-life of 2.3d by β-emission.

In this way, the breeder reactor produces energy as well as a fissionable material ^{239}Pu$_{93}$. Thus, we produce two generation reactors: a breeder reactor and a thermal reactor.

1.6.1 Thorium Fission

The atomic number of Thorium is 90. It occurs naturally. It has 90 protons and is a silvery element. It is weakly radioactive. All its isotopes are unstable except Thorium-232 which has 142 neutrons. It is expected to be present in the earth's crust 3–4 times more than Uranium. It is refined from monazite sands as a by-product in the extraction of rare earth metals. In India, there is a large deposit of Thorium and scientists are trying to produce fissionable material from thorium. It is expected that thorium might replace Uranium as a nuclear fuel in nuclear reactors. $^{232}Th_{90}$ is a fertile material from which $^{233}U_{92}$ can be generated. Thus like $^{238}U_{92}$ which is used to produce $^{239}Pu_{93}$, Thorium-232 can be used in a breeder reactor to produce Uranium-233. Hence, like Uranium-238, Thorium-232 can be used to develop two generation reactors. $^{232}Th_{90}$ when it captures neutrons (irrespective of whether in a fast breeder reactor or a thermal reactor) produces $^{233}Th_{90}$. This emits an electron and anti-neutrino (ν) by β-decay to become $^{233}U_{92}$ as shown in the decay scheme:

$$^{1}n_0 + {}^{232}Th_{90} \rightarrow {}^{233}Th_{90}^{\beta} \rightarrow {}^{233}Pa_{91}^{\beta} \rightarrow {}^{233}U_{92}$$

In future, we may see many Thorium-based breeders and thermal reactors in India and then shall not be dependent on other countries to get the fissionable Uranium-233 element.

1.6.2 Fusion Reaction

One of the disadvantages of a fission reaction is that it produces a large quantity of radioactive isotopes and some of them have a half-life of several years. Hence, the disposal of radioactive material is a problem with such type of reactor. Alternatively, H or He can also undergo a similar fission reaction synthesizing higher mass number nuclides (Fig. 1.1). The energy released is also almost the same as the uranium fission reaction. The advantage of this type of reaction is that it produces no radioactive isotopes, and it does not produce a mass number of atoms greater than 12. Since in such reactions, two atoms are combined or are fused together to produce an atom with higher binding energy, this reaction to distinguish it from the Uranium fission reaction is called the **Fusion** reaction. It is clear from Fig. 1.1 that $^{4}He_2$ has the maximum mass difference among the lighter atoms. This suggests that H is unstable in comparison to the He atom. In order to form one He atom, 4 H atoms are required. The quantification of mass difference can be obtained by considering the masses of H and He atoms

$$\text{Mass of } 4^{1}H_1 = 4 \times 1.00728 \text{ a.m.u.}$$
$$= 4.02912 \text{ a.m.u.}$$
$$\text{Mass of } {}^{4}He_2 \text{ atoms} = 4.00150 \text{ a.m.u.}$$

Mass difference or excess mass released by the formation of $^{4}He_2$ atom $= 0.02761$ a.m.u. $\times 931 = 25.7$ MeV. This means that when 4 atoms of H are fused to get He atoms, they release 25.7 MeV energy. Thus, the energy released in the H fusion reaction per four He is 25.7 MeV. A by-product of the H fusion reaction is the formation of a He atom.

Scientists are trying to create H fusion in the laboratory but it has not been very successful because one needs a temperature in the region of 10^8 K. It has been difficult to maintain such a high temperature for a longer period. This is a blessing in disguise. If scientists succeed in this fusion reaction, there would be a scarcity of water on the earth, because the source of H is water and in a fusion reaction, the by-product is He and other atoms and not water, as is the case if we burn H in the presence of oxygen. Hence, if such reactors succeeded, then there will be a scarcity of water which is the main material to sustain our lives on this planet. There are many speculations of the type of reactions which can take place in a H fusion reaction; one of the most accepted reactions as given here:

$$^1H_1 + {}^1H_1 \rightarrow {}^2D_1 + \beta + \mu$$
$$^2D_1 + {}^1H_1 \rightarrow {}^3He_2$$

or

$$^2D_1 + {}^2D_1 \rightarrow {}^4He_2 + \gamma$$

or

$$^3He_2 + {}^3He_2 \rightarrow {}^4He_2 + 2{}^1H_1$$

where $^2D_1$ is the isotope of H known as Deuterium with one neutron and one proton making mass equal to 2. Though the H fusion reaction has not been very successful by our scientists, this reaction is the main source of the generation of power from sunlight. It is now expected that H fusion is coming to near completion and the by-product of H fusion, i.e., He is now undergoing fusion reaction in the Sun. Since power produced by He fusion is more than that of H fusion, the Sun's temperature is increasing causing global warming.

1.7   Stability of Nucleon

We can get some useful information about the stability of naturally occurring nuclides by plotting a graph between neutron and proton numbers (Fig. 1.2). It is observed from this figure that the neutron–proton ratio for nuclides of a mass number less than 20 is approximately one and beyond this mass number, the ratio increases to almost 1.5. The ratio increases to counter the increase in repulsive forces between protons. As the number of protons in the nucleus increases, repulsive forces between them increase. This is compensated by a greater increase in the number of neutrons, and the neutron/proton ratio increases to about 1.5 or so for heavy nuclides.

A survey of stable nucleus reveals that the number of stable nuclei is large for those elements which possesses an even number of both neutrons and protons (162). There are very few stable nuclides, which possess odd numbers of neutrons and protons

Fig. 1.2 Neutron–Proton
ratio of naturally occurring
elements

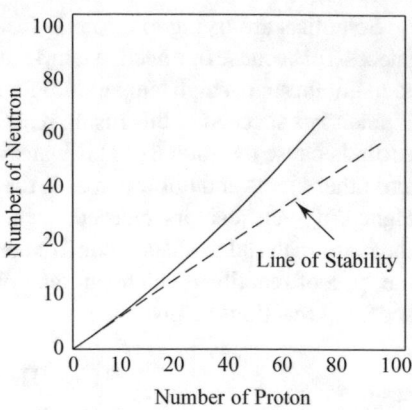

(4) and about 52–56 with even/odd number of nucleons (Table 1.1). This analysis seems to indicate that for forming a stable isotope, neutrons and protons prefer to form pairs with their own species but not with each other. As a result, most of the stable nuclides, e.g., $^4He_2$, $^{12}C_6$, $^{16}O_8$, $^{20}Ne_{10}$, and $^{26}Mg_{13}$ are of the even–even type. The nuclear stability chart is also available, which lists out all stable and radioactive isotopes (including man-made isotopes for each element of the periodic table) of naturally occurring nuclides; this chart (not shown here) will reveal that there are some elements which possess a large number of stable isotopes. These elements, interestingly enough, contain a specific number of protons, e.g., 2, 8, 20, 50, 82, and 126. Likewise, elements showing a large number of stable isotopes also contain the same number of neutrons, e.g., 2, 8, 20, 50, 82, and 126. These numbers, 2, 8, 20, 50, 82, and 126, are therefore called "**magic number**".

Furthermore, this nuclear chart reveals that among most elements, nuclides of mass numbers 9, 13, and 17 show a special feature. Isotopes of these mass numbers are unstable, especially when these nuclides contain one more neutron than protons, e.g., $^9Be_5$, $^{13}C_6$, and $^{17}O_8$ or when they contain one more proton than neutrons, e.g., $^9B_4$, $^{13}N_7$ and $^{17}F_9$. From the analysis of the stability of isotopes, it appears that the nuclei of these species contain pairs of neutrons and pairs of protons, each pair occupying a particular energy level. The odd nucleon may be occupying a higher

Table 1.1 Distribution of naturally occurring stable nuclides with various combinations of neutrons and protons

Number of protons	Even	Odd	Even	Odd
Number of neutrons	Even	Odd	Odd	Even
Number of stable nuclides	162	4	56	52

energy level and thus would have a tendency to lose one odd nucleon to become a stable isotope. Therefore, it seems reasonable to suppose that for the stability of these isotopes and the energy level occupied by odd protons in these nuclides must be higher than that of odd neutrons. These observations form the basis of the "**shell model**" of nucleus.

Summary

In this chapter, we have related the structure of the atom to explain the shell model, how the concept of binding energy is helpful in understanding the stability of nucleons, and about nuclear reactors (thermal and breeder reactors) dealing with fission reaction of uranium, thorium and fusion reaction with H. A discussion is made to explain how the ratio of protons and neutrons affects the stability of nucleons. The concept and application of a magic number are also touched upon.

Chapter 2
Radioactivity

2.1 Introduction

This chapter is devoted to covering various aspects of radioactivity. It encompasses properties of particulate (i.e., α, β, and positron) radiations and electromagnetic radiation, the law of radioactive decay, and the statistics of counting. The emission of radiations from a radioactive material comes from the fact that a radioactive isotope is unstable and it converts to a stable isotope by emitting radiations like α, β, positron, γ-rays, etc. In order to understand the nature of these radiations, it would be pertinent to recall atomic and nuclear structures in brief.

When atomic nuclei have a greater or lesser number of protons or neutrons than what is required for a stable nucleus, they rearrange their nucleons in order to achieve a more stable neutron/proton ratio; this rearrangement is independent of the chemical or physical state of the element. Sometimes, the nucleus may undergo a series of changes before it ultimately attains stability. The unstable nuclides are known as "**radioactive isotopes**" and the phenomenon of radiation emission is called "**radioactivity**". The method of rearrangement is called a "**decay process**", and is independent of past life or method used for the production of nuclei. Based on pure thermodynamic consideration of mass defect, radioactivity and possible types of radioactive decays are discussed in the forthcoming sections.

If mass defect is positive, then the process is said to be **exoergic decay** and if negative, then we call it **endoergic decay**. Based on mass defect and thermodynamic consideration, one can classify the radioactive nuclides' decay under three main headings.

2.2   Emission of Nuclear Particles

Radioactive atoms possess an excess of either protons or neutrons than required to be a stable isotope. The easiest way to remove excess neutrons or protons would seem to be by ejecting them directly from the nucleus. But in reality, such processes are very rare. Why? Since radioactive decay is a natural process, the unstable nuclei must be able to do this without receiving energy from outside. Neutron or proton emission is not possible unless energy equivalent to the binding energy of the ejected proton or neutron is put into the nucleus. For example, the average binding energy per nucleon of the last few nucleons for elements of mass number between 209 (i.e., ^{209}Bi) and 238 (i.e., ^{238}U) is about 5.5 MeV and for lower mass number elements, it is about 7–8 MeV. Hence, the acquisition of energy equivalent to this (i.e., 7–8 MeV) would be necessary for one nucleon to be expelled out of the nucleus. However, on the other hand, if two protons along with two neutrons are to be removed from the nucleus of a heavy element, then an energy equivalent to about 22 MeV would be required. This energy is less than the binding energy of an α-particle (i.e., about 28.28 MeV as calculated earlier). Thus, the ejection of an α-particle from the nucleus of a heavy element is exoergic and spontaneous, and α-decay is able to take place rather than decay by the emission of either protons or neutrons. Radioactive decay for nuclides of lower mass number such as, α-decay is usually not feasible, owing to higher binding energy of nucleons. In order to explain these two aspects of the decay process, let's take two examples: decay of ^{238}U and ^{39}K isotopes.

$$^{238}U_{92} \rightarrow ^{234}Th_{90} + ^{4}He_2$$
$$(238.05076 \text{ a.m.u.}) (234.043570 \text{ a.m.u.}) (4.002603 \text{ a.m.u.})$$

The difference in mass between the parent (^{238}U$_{92}$) nuclei and daughter nuclei (^{234}Th$_{90}$) is 0.0045 a.m.u. This excess mass is equivalent to 4.26 MeV. Therefore, it is possible for this isotope to decay by an α-decay process energetically. On the other hand, ^{39}K

$$^{39}K_{19} \rightarrow ^{35}Cl_{17} + ^{4}He_2$$
$$(38.971458 \text{ a.m.u.}) (34.9688545 \text{ a.m.u.}) (4.002603 \text{ a.m.u.})$$

cannot spontaneously undergo α-decay because the total mass of the two daughter nuclei (^{35}Cl$_{17}$ + ^{4}He$_2$) is heavier (i.e., 38.971458 a.m.u.) than the parent nuclei (^{39}K$_{19}$) (i.e., 34.9688545 a.m.u.), and so the available energy for transition is negative. These two examples suggest that on the basis of mass calculation, it is possible to theoretically predict the possibility of α-decay of any isotope.

The process of α-decay can be symbolically represented as

$$^{A}K_z \rightarrow {}^{A-4}Y_{z-2} + {}^{4}He_2$$

The daughter nuclei, i.e., the product nuclei are an isotope of the element with an atomic number two units lower than that of the parent.

In these calculations, the shock experienced by the parent nuclei when an α-particle leaves the nucleus is not considered. Since an α-particle has a mass and it leaves the nucleus with a velocity almost equal to the velocity of light, the parent nuclei must also experience a recoil. The energy of α-particles emitted should, thus, include the effect of recoil energy as well. Therefore, the actual energy of the α-particle is the difference between energy available for transition and recoil energy of the daughter nucleus. When an α-particle having a mass of 4.002603 a.m.u. leaves the nucleus, recoil is experienced by the nucleus. Hence, part of the energy is also shared in conserving the recoil energy. Considering these factors, the energy of an α-particle can be calculated from the expression:

$$E = E_\alpha \frac{M_d + M_\alpha}{M_d} \tag{2.1}$$

where

$E =$ Total energy available for α-decay (MeV) obtained from mass balance calculation.
$E_\alpha =$ Actual energy of an α-particle.
$M_\alpha =$ Mass of an α-particle.
$M_d =$ Mass of the daughter nucleus (Y).

Equation (2.1) can be used to calculate the actual energy of α-particles (E_α) emitted from any nucleus, provided the total energy involved in the decay process is known from the mass calculation or vice versa. For example, in the previous example of Uranium-238 decay, the total energy involved in the decay process was calculated to be 4.26 MeV. Therefore, considering the recoil energy, the α-particle energy should be 4.18 MeV, which is the experimentally observed value.

These calculations, thus, suggest that mass calculation can be used for getting an idea about the energy of α-particles expected to be emitted from a radioactive isotope. Moreover, it can also confirm the feasibility of α-decay by an isotope. The reason for not observing proton or neutron decay and observing α-decay can also be understood from such mass calculations. For the latter type of decay, the difference between the mass of the parent radioactive isotope and the daughter nuclei must be greater than the mass of He atom, i.e., 4.002603 a.m.u.

It is interesting to find out the type of decay possible if this mass difference is less than the mass of the He atom. It will be seen in the forthcoming discussions that if an isotope contains an excess proton than needed for its stability, the decay normally takes place by the conversion of the excess proton into a neutron or an excess neutron into a proton, as the case may be. This type of decay is discussed here.

2.3 Interconversion of Nucleons Within the Nucleus

2.3.1 Conversion of Neutrons into Protons

An unstable nucleus that has excess neutron, but not enough to combine with proton in pairs and emit α-particles and become stable; are made stable by converting a neutron into a proton. Nuclear electron known as a β-particle (to maintain conservation of electric charge) along with electromagnetic radiation known as a neutrino is emitted. When a neutron is converted into a proton, some excess mass (i.e., the mass difference between the parent and daughter + β-particle) may still be left. This excess mass may not be large enough to allow another neutron to convert into a proton. Therefore, this un-utilized excess mass is converted into electromagnetic radiation. This type of electromagnetic radiation is called a neutrino. In other words, β-particles together with a neutrino are emitted to conserve excess mass left in the nucleus after one neutron has been converted into a proton.

The energy available for β-transition (i.e., mass excess) is, thus, shared between a neutrino and a β-particle. In α-decay, there is no such sharing of excess mass, hence, the α-particle is emitted from the nucleus with a specific kinetic energy whereas, in β-decay, since the excess mass (i.e., the energy converted out of the excess mass) is shared between a neutrino and a β-particle, emitted radiation does not possess fixed kinetic energy. As a result of this sharing, in β-decay, the energy of the β-particle does not have a specific value, instead, it results in a continuous energy spectrum. The examination of a typical spectrum of the energy of β-particles, as shown in Fig. 2.1, suggests that β-particles can possess any value of energy from zero to maximum energy.

Maximum energy is expressed as E_{max}. But the intensity of β-particles (i.e., number of β-particles of energy E_{max} per unit time) is very less. On the other hand, β-particles with the number of β-particle per unit time (i.e., highest intensity) are those which have energy approximately equal to E_{max}. β-decay thus can be written as follows:

Fig. 2.1 β-spectrum of a radioactive isotope

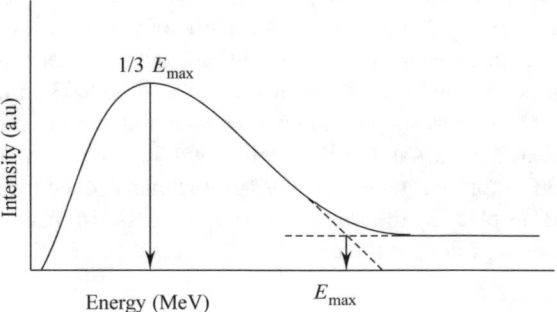

$$n \rightarrow p + \beta^- + \mu$$

2.3.2 Conversion of Proton to Neutron

Like the conversion of an excess neutron to a proton, a nucleus with an excess proton (i.e., excess than needed for its stability) can approach stability by converting a proton into a neutron by capturing one of the nearest orbital electrons. This type of decay is known as **electron capture (E.C.)**. Unlike β-decay, the excess mass (i.e., energy converted from the excess mass) left even after converting the proton by capturing the orbital electron is converted into a neutrino (μ). This type of decay can be expressed as follows:

$$p + e^- \rightarrow n + \mu.$$

If for some reasons, electron capture is not possible, the proton can be converted into a neutron by the emission of a positron (β^+); this type of decay is known as **positron decay**. In this type of decay also the neutrino is ejected for the same reason as expressed earlier. This decay can be expressed as follows:

$$p \rightarrow n + \beta^+ + \mu.$$

A positron is a positively charged electron designated by β^+. The question is: which of these two types of decays will an isotope having an excess proton prefer? This can be found out by considering mass excess in each decay process. In decay by positron emission, the atomic number of the atom is decreased by one unit, and one of the orbital electrons becomes superfluous:

$$\begin{array}{ccc} ^A K_Z & \rightarrow & ^A Y_{Z-1} & + \beta^+ + e + \text{decay enery} \\ (Z \text{ electrons}) & & (Z-1, \text{ electrons}) \end{array}$$

If atomic masses are used for the calculation of positron decay energy, the mass of this superfluous electron must be considered. Thus,

$$(^A X_Z - {}^A Y_{Z-1}) - (\beta^+ + e^-) = \text{positron decay energy}$$

This calculation suggests that for positron emission to occur spontaneously, the difference in mass between parent and daughter atoms must be greater than the combined mass of the positron and electron. That is,

$$(^A X_Z - {}^A Y_{Z-1}) > (\beta^+ + e^-).$$

This may be illustrated by considering an example of Sodium-22 decay.

$$^{22}\mathrm{Na}_{11} \rightarrow \beta^+ + {}^{22}\mathrm{Ne}_{10}$$

In this example, Sodium-22 has 11 electrons while Neon has only 10 electrons, which means that the mass of one electron should be considered in the calculation. It is important to realize that a positron (i.e., β^+) has been evolved from the nucleus due to the conversion of the proton into a neutron and is not due to orbital electrons of Sodium and Neon.

Mass of Sodium-22	= 21.994435 a.m.u
Mass of Neon-22	= 21.9913845 a.m.u
Mass difference	= 0.0030505 a.m.u
Energy equivalent	= 0.0030505 a.m.u. $\times$ 931.5 MeV
	= 2.84 MeV
Mass of one electron	= 9.10939×10^{-31} kg
	= 0.511 MeV
because 1 a.m.u.	= 1.66054×10^{-27} kg
Thus mass of two electrons	= 2×0.511 MeV, i.e., $(e^- + e^+) = 1.022$ MeV

This calculation suggests that since energy equivalent to mass difference ($^{22}\mathrm{Na}_{11} - {}^{22}\mathrm{Ne}_{10} = 2.84$ MeV) is greater than the sum total of electron and positron energies ($e^- + e^+ = 1.022$ MeV), positron emission is energetically possible. This is also the case with Sodium-22; it decays by positron emission. On the other hand, decay by electron capture simply involves the transfer of an orbital electron to the nucleus (i.e., it is not lost from the system).

Thus, for an electron capture decay, mass of $^A\mathrm{X}_Z$-mass of $^A\mathrm{Y}_{Z-1}$ must be greater than zero. In other words, for electron capture decay, it is only necessary for the atom $^A\mathrm{X}_Z$ to be heavier than $^A\mathrm{Y}_{Z-1}$.

The unstable nuclei $^{55}\mathrm{Fe}_{26}$ that decay to $^{55}\mathrm{Mn}_{25}$ by electron capture can be considered as an example to illustrate this phenomenon.

Mass of Iron-55	= 54.9383024 a.m.u
Mass of Manganese-55	= 54.9380536 a.m.u
Mass difference	= 0.0002488 a.m.u
Energy equivalent	= 931.5×0.0002488 MeV = 0.231 MeV

Thus, Iron-55 can decay only by electron capture, which is observed in its decay. Hence as a general rule, it can be remembered that if mass difference of ($^A\mathrm{X}_Z - {}^A\mathrm{Y}_{Z-1}$) is greater than the energy equivalent to ($e^- + e^+$) = 1.022 MeV, positron emission as well as electron capture are possible; but if this energy is less than 1.022 MeV, only electron capture is possible. However, it is observed that among the lighter elements, positron emission and electron capture are approximately equally probable, but as the atomic number of nuclei increases, the orbital electrons are drawn toward the nucleus and the probability of electron capture increases. Positron emission is not usually observed in heavy elements.

The detection of an isotope decaying by electron capture becomes a difficult task because excess energy is released in the form of X-rays. Since electron capture occurs with electrons of the innermost orbital, the vacancy created in this orbit is filled by dropping electrons from the next neighboring orbital. The difference in the energies of the two orbitals is emitted as X-rays. This process leads to emissions of X-rays of different energies, because the process of filling of electrons from neighboring orbitals continues in the atom, until an electron of the uppermost orbital has adjusted by the process of filling the vacancy created in its next orbital. Therefore, the energy of X-rays due to electron capture and filling of vacancy may not always be much different from each other. Thus, the identification of an isotope, which decays by electron capture, becomes a difficult task.

Very useful information concerning β-decay can be drawn by calculating binding energy of various isobaric elements (especially for mass number greater than 80), using an empirical equation mentioned below:

Binding Energy

$$(\text{MeV}) = 14.0\,A - 19.3\frac{A - 2Z}{A} - 0.585\frac{z}{A^{\frac{1}{3}}} - 13.05\,A^{\frac{2}{3}} \pm \frac{125}{A} \qquad (2.2)$$

where A is mass number and Z is proton number. This equation has been developed by considering various factors, which may have an effect on the binding energy of a nucleus. In this equation, the first term accounts for short range nuclear forces between two nuclei. The binding energy of the nucleus decreases accordingly with the extra number of neutrons, hence the second term is subtracted from the binding energy value. Increase in repulsive energy between protons is inversely related to distance, i.e., the radius of the nucleus. Accordingly, the binding energy decreases. This accounts for the third term. With increase in radius of the nucleus, $A^{1/3}$, which is proportional to the surface area of nucleus, also increases. In other words, the number of less tightly bound nucleons near the surface increases accordingly, followed by an expected decrease in binding energy. This accounts for the 4th term. The last term considers even–odd character of the number of protons and neutrons present in the nucleus. This value is positive for even–even nuclei, negative for odd–odd, and zero for odd–even nuclei.

This equation has been used by the author to calculate the binding energies of isobaric nuclides of mass number 144. The results are plotted against the neutron number, giving two parabolas (Fig. 2.2), for odd and even number of neutrons. The advantage of this calculation is that one can easily decide which of the isobaric nuclides of 144 mass number will decay by β-decay and which will decay by positron or electron capture type decay. Moreover, an idea about the magnitude of instability can also be realized from this curve. For example, $^{144}Xe_{54}$ is energetically very unstable and will have the tendency to decay to $^{144}Cs_{55}$ by β-decay process.

Elements lying on the left-hand side of the curve are unstable to β-decay, while those present at the beginning of the two parabolas (i.e., before Cerium-144) show a very sharp change in binding energy; hence the energy available for decay is large. In other words, these isotopes will decay with the emission of the β-particle having the maximum E_{max} value as compared to nuclides like $^{144}La_{57}$, which will decay to $^{144}Ce_{58}$ with lower E_{max} energy. Nuclides on the right-hand side of the parabolas are deficient in neutrons; they decay by positron emission or electron capture processes.

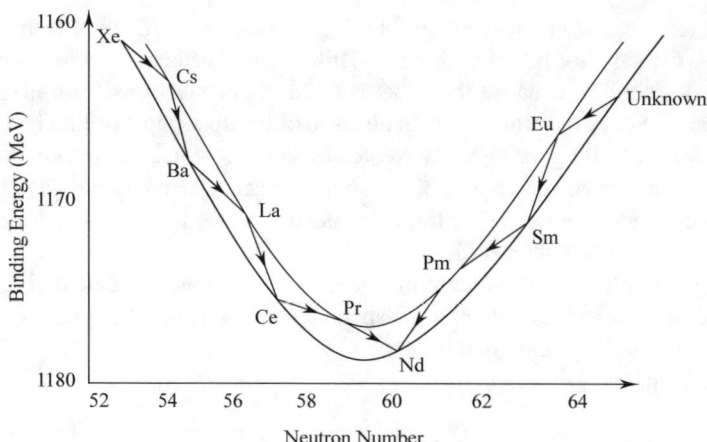

Fig. 2.2 Binding energy curve versus neutron numbers, calculated from Eq. (2.2) for mass number 144

The elements near Neodymium-144 have binding energy values very close to each other and the decay energies are, therefore, low. Neodymium-144 according to the curves should be stable as it has the highest binding energy, and Samarium-144 should be unstable because its binding energy is lower than that of Promethium-144. In fact, Neodymium-144 is stable toward β-decay and Samarium-144 is also stable, whereas Promethium-144 is not. Presumably, this is because Samarium-144 has the magic number of 82 neutrons, which is not taken into account in the binding energy equation.

2.4   Transition Between Nuclear Energy Levels Without Emission or Interconversion of Nucleons

In the previous discussions, we considered the mass difference between parent (unstable isotope) and daughter nuclei (isotope formed after the decay), to establish the nature of decay. During the decay of the isotope, it may happen that the product still contains some excess energy, which is not enough to carry out any further decay processes, by any of the methods discussed earlier. This excess energy leaves the daughter isotope in a somewhat excited state. The excited daughter nuclei come down to a stable state (often called the ground state) by emitting excess energy in the form of electromagnetic radiation. The magnitude of electromagnetic energy depends on the difference in energy between the excited state and expected ground state energy of the corresponding stable isotope. The electromagnetic radiation is known as γ-radiation. That is,

$$E_\gamma = E_1 - E_2 = h\nu$$

where

$E_\gamma =$ Energy of electromagnetic radiation
$E_1 =$ Energy of excited state
$E_2 =$ Energy of ground state
$=$ (or lower excited state than E_1)
$h =$ Plank's constant
$\nu =$ Frequency of radiation.

While the emission of γ-radiation usually takes place very rapidly, there are occasions when the excited nuclei take a very long time for such emission. Such a state of nuclei is called as **metastable** state and decays slowly. The decay process of ^{137m}Ba is one example of such decay process. This type of decay is known as an **isomeric transition (I.T.)**.

$$^{137m}\text{Ba} \xrightarrow{\text{2.4 min}} {}^{137}\text{Ba}.$$

Sometimes excess nuclear energy, instead of being emitted as electromagnetic radiation, is transferred to one of the orbital electrons, usually in the K-shell. This absorption of energy causes the ejection of electrons from the atom with energy equal to the difference between decay energy and binding energy of the concerned electron. This type of decay is known as **internal conversion**. The ejected electrons are monoenergetic, in contrast to the heteroenergetic β-particles. An X-ray radiation is also emitted, as the vacancy in K-shell is filled by the electrons falling from higher orbitals. This type of emission may lead to X-ray emission of different energies due to a series of falls of electrons from various higher orbitals to next lower energy orbitals and finally to the K-shell orbital. Approximately, 12% of the disintegration of Barium-137m is by internal conversion (Figs. 2.3 and 2.4). Along with γ-rays (Fig. 2.3B), some X-ray (Fig. 2.3A) is also emitted. Decay by an internal conversion may have interesting consequences if the atom concerned is covalently bonded to another atom, as one of the electrons forming the bond may fill the vacancy in the inner orbital. In this case, the chemical bond would be ruptured.

These discussions thus also suggest that practically in every decay process, irrespective of its kind, there is a possibility of γ-ray as well as X-ray emission. Because it is very unlikely that after the decay of the radioactive isotope by any of the processes discussed earlier, there would not be some excess mass left (i.e., mass of the isotope being more than required for becoming stable isotope) with the daughter nuclei, this excess mass is not enough to emit, for example, α- or β-particles and hence the excess mass is emitted in the form of either γ-rays or X-rays.

In fact, the emission of γ-ray is advantageous to us, because the energy of γ-ray is very specific for a radioactive isotope and thus can help to detect the presence of isotope in the sample by measuring γ-rays (Fig. 2.4C) rather than β-particle (Fig. 2.4B) or X-rays (Fig. 2.4A).

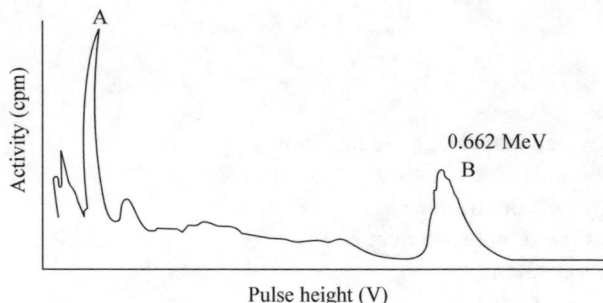

Fig. 2.3 γ-ray (0.662 MeV) spectrum of Cesium-137/Barium-137 m: (A) peak due to X-ray and (B) for γ-ray

Fig. 2.4 β-spectrum of Cesium-137/Barium-137 m, showing (A) peak due to X-ray, (B) a continuous β-particle spectrum, and (C) peak due to internal conversion electron spectrum

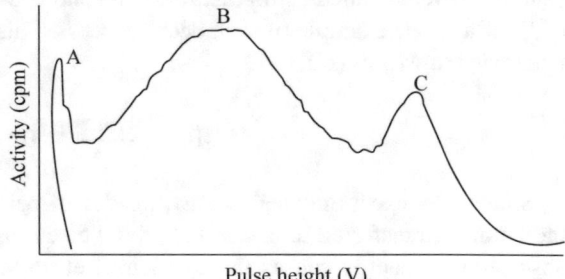

2.5 Natural Radioactive Series

There are some radioactive isotopes which decay to another radioactive isotope which decays further to another radioactive isotope. This type of decay continues till it comes to a stable isotope. During this series of decays, the radioactive isotopes decay by α-particles or β-particles. Moreover, the half-life of decays varies from a few seconds to a few years. It is interesting to note that the Mass number of all series is divisible by digit 4. Therefore, these series are also expressed as four times an appropriate integer $(n) + 1$. For example,

- $(4n)$ as the Thorium series starting with Thorium-232;
- $(4n + 1)$ is Neptunium-237 (Fig. 2.5);
- $(4n + 2)$ is the Uranium series with Uranium-238 (Fig. 2.6);
- $(4n + 3)$ is the Actinium series with Actinium-235.

However, the natural form of $(4n + 1)$ has not been found, though it has been synthesized artificially. The reason could be the very short-lived half-lives of some of the isotopes of the $(4n + 1)$ series.

Thus scientists tried to synthesize artificially elements of this series.

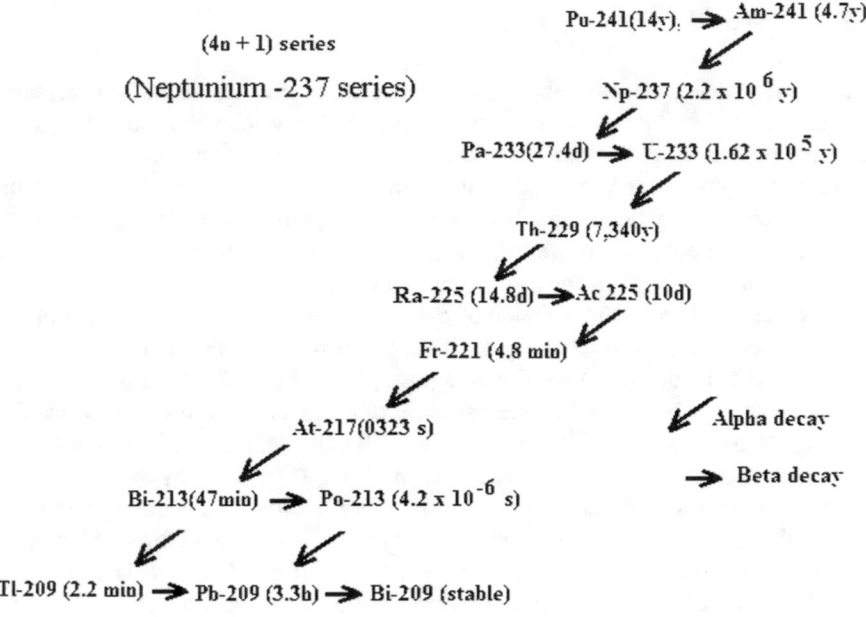

Fig. 2.5 Decays of $(4n + 1)$ series of Neptunium-237

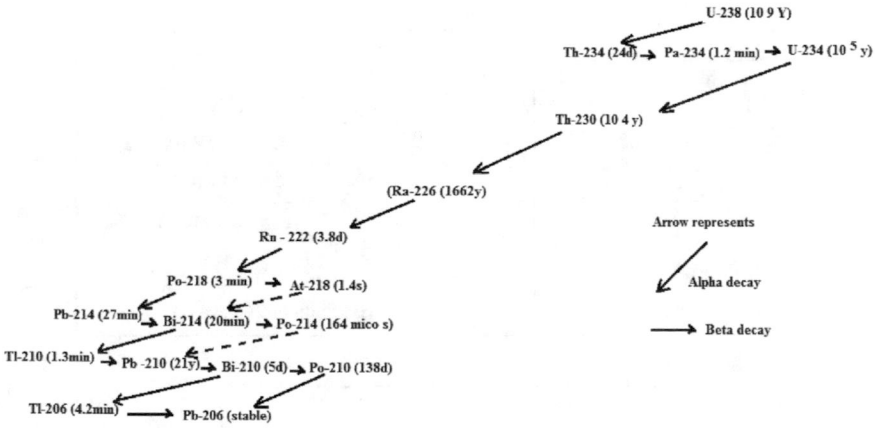

Fig. 2.6 Decays of $(4n + 2)$ series of Uranium-238

2.6   Decay Scheme

The nature of decay of many radioactive isotopes has been experimentally determined and is available in the form of a decay scheme. A decay scheme is an energy level diagram representing the manner in which a radioactive nucleus decays to another species. This decay scheme helps to understand the nature of decay, energy of the radiation emitted by the isotope, % of decay in the particular mode (because many radioactive isotopes decay by more than one mode of decay), half- life of the isotope, etc. Some of the decay schemes are illustrated in Fig. 2.7.

As discussed earlier, the emission of β-particles often leaves the daughter nuclei in an excited state; the surplus energy is then lost by the emission of γ-rays. The decay of $^3_1$H is a good example of such a pure β-decay (Fig. 2.7A). Sometimes, the ground state may be reached by the emission of more than one γ-ray in successive steps, as seen in case of the ^{24}Na decay (Fig. 2.7B). In addition, a radioactive isotope may become a stable isotope by the emission of one β-decay or by the emission of several β-particles of different energies, one after another in a sequential fashion. Alternatively, it may decay by the emission of various β-particles of different energies simultaneously (decay of ^{90}Sr is a good example of this type of decay (Fig. 2.7C). The probability of decay by each of these processes may either be the same or different. Each of these decay processes may or may not be followed by the emission of γ-rays.

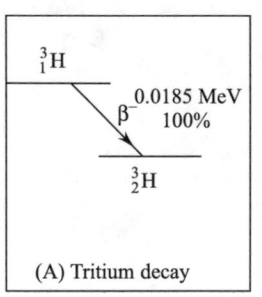

(A) Tritium decay

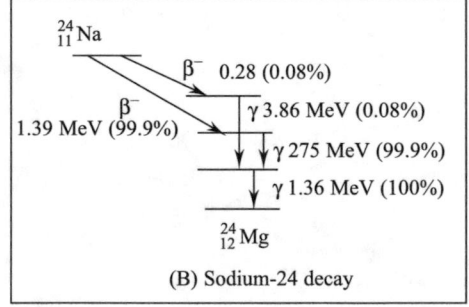

(B) Sodium-24 decay

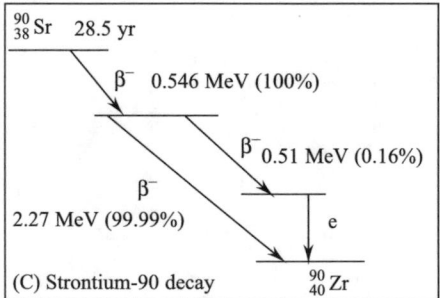

(C) Strontium-90 decay

Fig. 2.7   Decay schemes of (A) Tritium, (B) Sodium-24, and (C) Strontium-90

The γ-ray emission will depend upon the amount of excess mass (i.e., whether the mass of the isotope is equal to the mass of the stable isotope) remaining with the daughter isotope after each decay. Thus, we see that the decay process of an isotope can be very complex. The decay of ^{90}Sr to either ground state ^{90}Sr via various other excited states (^{90}Y is the best example to illustrate this process (Fig. 2.7C). There are, of course, many other decay schemes both simple and extremely complex.

2.6.1 Rate of Decay

Unstable nuclides can approach stability by a suitable mode of decay, but none of these decay processes are instantaneous. The unstable nuclei may require a millionth of a second or millions of years to decay to a more stable one. The rate of decay can be expressed using the differential form:

$$\frac{dN}{dt} \propto N = -\lambda N \tag{2.3}$$

in which "N" is the number of radioactive atoms dN is the change in the number of atoms and dt is the change in time. By introducing a proportionality constant λ, and indicating a decrease in activity with a negative sign, this equation becomes equal to the product of λN. In other words, radioactive decay follows the law of first-order kinetics. Integration of Eq. (2.3) gives the following solution:

$$N = N_0 e^{-\lambda t} \tag{2.4}$$

in which N_0 is the number of radioactive atoms at some reference time, "N" the number of radioactive atoms at time "t", and λ is the decay constant. λ is a fundamental constant for each nucleus and has the dimension of reciprocal time. It is a measure of the probability that a given single nucleus will decay within unit time.

It may not always be possible to measure absolute values of N or N_0. However, each atom of the radioactive isotope emits one radiation per decay of the atom. If the radiations emitted by atoms could be measured for some period and the value is converted to a unit "**number of radiations recorded per unit time**", i.e., **activity** (expressed in terms of disintegration rate or counting rate), the Eq. (2.4) can then be expressed as

$$A = A_0 e^{-\lambda t} \tag{2.5}$$

in which "A" is the activity at time "t" (i.e., counts per unit time) and A_0 is activity at some reference time (i.e., counts per unit time). This relationship can be used to determine the half-life of the radioactive isotope.

2.6.2 Half-Life

The radioactive nuclides are generally described by the term half-life, i.e., time required by a radioactive nuclei to disintegrate to half of its initial activity. Half-life is a characteristic of a radioactive species, and it is very unlikely that any two nuclides will have exactly the same value. The half-life is given by substituting the value of $A = A_o/2$ in Eq. (2.5), then it is

$$\frac{N_0}{2} = N_0 \exp(-\lambda t_{1/2}).$$

Thus

$$t_{1/2} = \frac{\ln 2}{\lambda} = \frac{0.693}{\lambda} \tag{2.6}$$

where $t_{1/2}$ is the half-life of the radioactive material. These laws, which give rise to such concepts as half-life, decay constant, etc., are statistical in nature because decay is a random process, and they are only valid when a large number of radioactive atoms are under consideration. However, determining the half-life of an isotope helps in the identification of the isotope because no two radioactive isotopes have the same half-life value.

2.6.3 Radioactive Equilibrium

In some cases, one type of radioactive isotope (A) decays to another isotope (B), which is also radioactive. That is,

$$A \xrightarrow{\lambda_A} B \xrightarrow{\lambda_B} C \text{ stable)}$$

For example, Strontium-90 decays into another radioactive species Yttrium-90 (Fig. 2.7C). With such type of materials, there is a possibility of the formation of equilibrium between the parent (A) and daughter (B) radioactive isotopes. The nature of equilibrium depends upon the half-life of the two isotopes. The development of a mathematical model for such equilibrium condition is attempted here. Suppose that atom "A" at time $t = 0$ possesses N_{A0} number of atoms with its decay constant λ_A, and decay constant of atom "B" is λ_B. Then after decaying of atom "A" for a period "t", the quantity of N_{At} can be given by

$$N_{At} = N_{A0} \times e^{\lambda_A t} \tag{2.7}$$

and

$$\frac{dN_A}{dt} = -\lambda_A \times N_A \tag{2.8}$$

Likewise, at time "t" the number of atoms of "B" being created per second from "A" is $\lambda_A N_A$ (i.e., number of atom A decaying into atom B) while the number of atoms of "B" disintegrating, in turn, is $\lambda_B N_B$. The rate at which the number of daughter atoms builds up is equal to the rate of their formation from the decaying parent atoms less than the rate of their own decay. Therefore, at such times "t" can be stated as

$$\frac{dN_B}{dt} = N_A \lambda_A - \lambda_B N_B \tag{2.9}$$

If we substitute the value of N_A from Eq. (2.7) into Eq. (2.9), we obtain

$$\frac{dN_B}{dt} = N_{A0} \lambda_A e^{-\lambda_A t} - \lambda_B N_B \tag{2.10}$$

This equation can be solved after integration to get

$$N_B = \frac{\lambda_A}{\lambda_B - \lambda_A} N_{A0} \left(e^{-\lambda_A t} - \lambda_B t \right) + N_{B0} \times e^{-\lambda_B t} \tag{2.11}$$

On the right-hand side of Eq. (2.11), the first term gives the growth of the daughter from the parent and its decay, while the second term represents the concentration of the daughter present initially at time $t = 0$. If N_{B0} is zero at both $t = 0$ and $t = \infty$, which is to be expected for a radioactive daughter (B) growing from a radioactive parent (A), then

$$N_B = \frac{\lambda_A}{\lambda_B - \lambda_A} N_{A0} \left(e^{-\lambda_A t} - \lambda_B t \right) \tag{2.12}$$

It is also apparent that if $N_B = 0$ for $t = 0$ and for $t = \infty$, then N_B must also grow to a maximum at some intermediate time t_m. This can be derived from Eq. (2.12) by determining the time at which N_B is maximum. This can be done by differentiating Eq. (2.12) and setting at $t = t_{max}$

$$\frac{dN_B}{dt} = 0$$

Then,

$$t_m = \frac{\ln \left(\frac{\lambda_B}{\lambda_A} \right)}{\lambda_B - \lambda_A} \tag{2.13}$$

that is, when $t = t_m$, we have from Eq. (2.9):

$$\lambda_A \times N_A - \lambda_B \times N_B = \frac{dN_B}{dt} = 0 \tag{2.14}$$

Therefore, whatever be the relative half-lives of the parent or daughter, their activities are equal at a time t_m, that is, when the activity of the daughter reaches a maximum value.

A detailed examination of Eq. (2.12) suggests that there can be three variations of this equation, which are of importance. Two of them are for cases where the parent is longer lived than the daughter (i.e., either $\lambda_B > \lambda_A$ or $\lambda_B \gg \lambda_A$), and the third case is when the daughter product is longer lived than its radioactive parent (i.e., $\lambda_A > \lambda_B$). These cases we shall now deal with separately.

When $\lambda_B \gg \lambda_A$

Under this condition, Eq. (2.12) can be rearranged by ignoring λ_A into the following form:

$$N_B \times \lambda_B = \lambda_A \times N_{A0}(1 - e^{-\lambda_B t}) \tag{2.15}$$

One interesting feature comes out from this equation, i.e., if the mixture (i.e., atoms A and B) is allowed to grow together for a period 6–7 times the half-life of the daughter B, after making sure that at $t = 0$ atom B was absent completely, the exponential term becomes almost zero (i.e., $e^{-\lambda_B t}$) and the final equation takes a form:

$$N_B \times \lambda_B = \lambda_A \times N_{A0} \tag{2.16}$$

Thus for $\lambda_B \gg \lambda_A$, we see that the daughter B soon decays with the half-life of the parent A and their activities are equal. The equilibrium represented by the above equation is called **secular equilibrium**. Examples of secular equilibrium are provided by the radioactive decay of Radium-226 ($t_{1/2} = 1600$ years) to Radon-222 ($t_{1/2} = 3.82$ days), Strontium-90 ($t_{1/2} = 29.12$ years) to Yttrium-90 ($t_{1/2} = 64.0$ hours), or Cesium-137 ($t_{1/2} = 30y$) decaying to Barium-137 ($t_{1/2} = 2.6$ min). All these members of the former family emit α-particles, and the latter emit negative β-particles. The growth of Barium-137 activity from freshly separated Cesium-137 is illustrated in Fig. 2.8.

When $\lambda_B > \lambda_A$

The second case is when $\lambda_B > \lambda_A$ but not so great that λ_A could be neglected. However, for larger time "t" λ_B could be neglected as compared to λ_A. Under this condition,

Equation (2.12) can be rewritten as

$$N_B = \frac{\lambda_A}{\lambda_B - \lambda_A} N_{A0}(e^{-\lambda_A t}) \tag{2.17}$$

Since

Fig. 2.8 Growth of
Radon-222 activity from
freshly separated
Radium-226

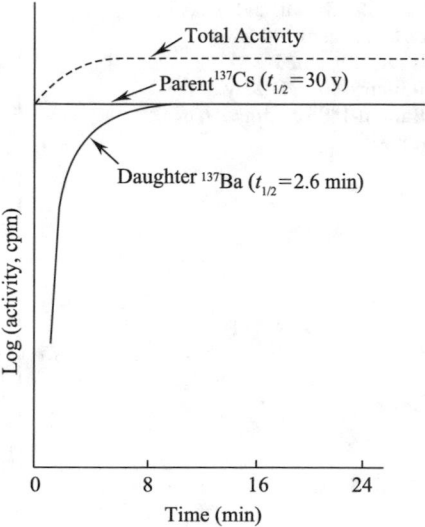

$$N_A = N_{A0} \times e^{-\lambda_A t}$$

we obtain

$$\frac{N_B}{N_A} = \frac{\lambda_A}{\lambda_B - \lambda_A} \qquad (2.18)$$

From Eq. (2.18), it is evident that the ratio of the number of daughter atoms to parent atoms becomes constant, and the rate of decay of the daughter is now determined by the rate of decay of the parent. This kind of equilibrium is known as **transient equilibrium**. An example of such case is shown in Fig. 2.9 for Barium-140 ($t_{1/2} = 12.74$ days), which decays to Lanthanum-140 ($t_{1/2} = 40.272$ hours). The condition of transient equilibrium gradually changes into secular equilibrium, as λ_B becomes larger as compared to λ_A.

When $\lambda_A > \lambda_B$

The third type of case is when $\lambda_A > \lambda_B$. Under this condition, Eq. (2.12) reduces to

$$N_B = N_{A0}(e^{-\lambda_B t}) \qquad (2.19)$$

This means that the N_{A0} (i.e., the parent atoms) very rapidly decays to an equal number of daughter atoms, which in turn, decay at a rate characteristic of the daughter. This is a case of no equilibrium.

In this case, the activity decays with a half-life of atom B, and the activity of parent atom A decreases more rapidly with its shorter life. An example of this case is the decay of Polonium-218 ($t_{1/2} = 3.0$ min) to Lead-214 ($t_{1/2} = 26.8$ min). A schematic graph of such type of decay is shown in Fig. 2.10. Under this case, there

Fig. 2.9 Growth and decay
of Lanthanum-140 from
freshly separated
Barium-140 and decay of the
Barium-140 as a function of
time

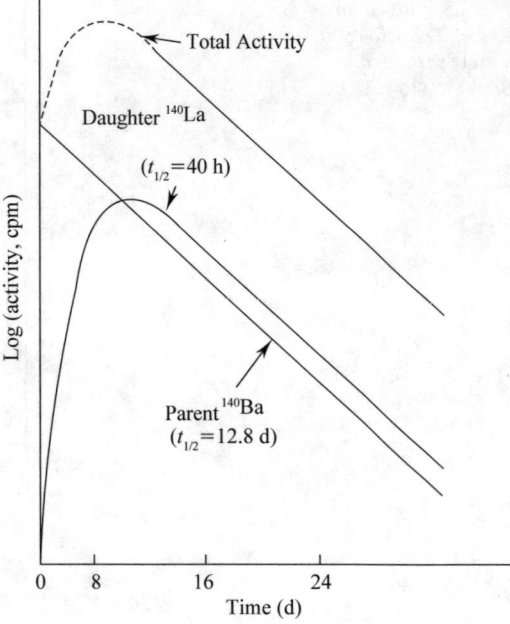

Fig. 2.10 Decay of
Polonium-218 to Lead-214

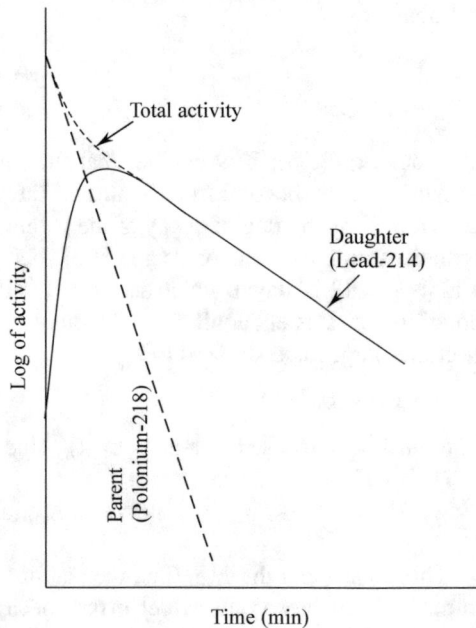

is no equilibrium. The advantage of these types of equilibrium is that one can use such conditions for measuring activities of radioactive materials, especially when both the parent and daughter are radioactive materials. This aspect will be discussed in a later chapter.

Summary

In this chapter, we discussed in detail all aspects of radioactivity, especially to predict the decay of any radioactive material by the calculation based on mass defect. Why a particular radioactive isotope decays only by a specific modes of decay, what are the possible types of decays—these are also touched upon. How the naturally occurring radioactive series and their classifications have helped to predict the existence of some isotopes which we do not get these days? The decay scheme of some specific examples is discussed to highlight its importance. The advantages of knowing the half-life of the isotopes, mathematical interpretation of different types of equilibrium, and their advantages are also dealt with.

Chapter 3
Nuclear Reaction

3.1 Introduction

With a material like ore, the nature of constituents present in it is not only unknown but is also present in nano-grams. Most of the conventional analytical techniques become unsuitable in analysis for such materials. The constituents of such sample can easily be estimated by a radiochemical technique. The sample is allowed to undergo nuclear reaction with nucleons of suitable energy to convert the stable isotopes present in a sample into corresponding radioactive isotopes. The identification of the isotopes and the estimation of the amount of corresponding isotopes produced help to conclude accurately the type of element and its quantity present in the original sample. Before discussing various radiochemical techniques, it may be appropriate to know about the nuclear reactions and factors, which control the formation of products from the corresponding nuclear reactions. In this chapter, we shall devote our attention to detail the nuclear reactions under separate headings.

3.2 Reactions Initiated by Charged Particles

Whenever, a charged particle approaches a nucleus, it experiences a Coulombic repulsive force and is deflected from its path. Thus, charged particles such as protons, doubly charged helium ions, and deuterons can approach a nucleus sufficiently closely to bring about a nuclear reaction, provided they have sufficient energy to overcome the Coulombic repulsive force. This is achieved by accelerating charged particles. For this purpose, an accelerating machine, such as a cyclotron is used. In this machine, charged ions are allowed to move in a circular long path under an increasing electric field. The longer the distance and the bigger the diameter of the cyclotron, the more is the energy acquired by the charged ions. For example,

© The Author(s), under exclusive license to Springer Nature Switzerland AG 2021
M. Sharon and M. Sharon, *Nuclear Chemistry*,
https://doi.org/10.1007/978-3-030-62018-9_3

Astatine-211 can be synthesized by bombarding Bismuth-209 with accelerated helium nuclei (32 MeV), from a 60 inch cyclotron.

$$^{209}\text{Bi}_{83} + {}^{4}\text{He}_2 \rightarrow {}^{211}\text{At}_{85} + 2\,{}^{1}n_0$$

A short description of this reaction is often given as $^{209}\text{Bi}(\alpha, n)^{211}\text{At}$. The chemical identity of the product may be established either by the separation of the constituent from the target element or by studying the chemistry of the expected products (i.e., Astatine) by a suitable radiochemical procedure. In this nuclear reaction, Astatine-211 from Bismuth-209 is separated by the volatilization process. Alternatively, the product may be isolated from the target material by using a chemically similar element to Astatine (i.e., chemically similar to the product of the reaction). For example, iodine which has a chemical property similar to Astatine can be used as a carrier. The usefulness of carrier for such application will be dealt with later.

Another example of a nuclear reaction induced by accelerated doubly charged helium ions is

$$^{239}\text{Pu}_{94} + {}^{4}\text{He}_2 \rightarrow {}^{242}\text{Cm}_{96} + {}^{1}n_0$$

The short description of this reaction can be written as $^{239}\text{Pu}(\alpha, n)^{242}\text{Cm}$. The product may be separated and its chemical identity recognized by its behavior on an ion-exchange column. Details of the ion-exchange technique will be discussed later.

From these examples, it is observed that the products formed by nuclear reactions involved with charged particles are usually chemically different from the target, e.g., both Bismuth-209 and Pu-239 give products of different elements ($^{211}\text{At}_{85}$ and $^{242}\text{Cm}_{96}$). The chemical separation of products from the target material thus becomes simpler for such type of nuclear reactions.

3.3   Reactions Initiated by Uncharged Particles

Neutron is the only uncharged particle that has a practical application in nuclear reaction and has the advantage that it does not require to overcome the Coulombic repulsive force of the nucleus. Broadly speaking, two types of nuclear reactions can be initiated with uncharged particle:

(i) **With energetic neutron** energy of interacting neutron may be enough to eject one or more than one nucleons from the nucleus (or)
(ii) **With thermal neutron** possessing energy in the vicinity of 0.03–0.1 eV (known as **thermal neutron**).

In the latter type, γ-emission is associated with the nuclear reaction. The energy of neutrons obtained from a neutron generator is in the range of a few MeV. Hence, for the former type of nuclear reactions, neutrons from the neutron generator can be directly used for carrying out the nuclear reaction. While in thermal neutron reaction, the energy of a neutron is slowed down to the energy of 0.03–0.1 eV.

The energy of a fast neutron is lowered by allowing it to interact with materials like pure graphite (known as **moderator**). This material instead of absorbing allows the neutron to undergo several collisions. In each collision, the neutron loses its kinetic energy and finally after several collisions, it becomes a thermal neutron.

3.3.1 Thermal Neutron Reaction

The most common type of nuclear reaction induced by a thermal neutron is the radioactive capture process, represented by (n, γ) in which a slow neutron is absorbed by the target. The excess energy (i.e., the difference between the energy of the interacting neutron and energy of the target nucleus) is emitted in the form of radiation. The product of such a nuclear reaction is the isotope of the target material with its mass number one unit higher. For example, thermonuclear reaction with Chlorine-35 present in $Na^{35}Cl$ will give an isotope of Chlorine-36 as $Na^{36}Cl$.

3.3.2 Fast Neutron Reaction

With fast neutrons, that is to say, with neutrons possessing energies in the range of 1–2 MeV, (n, p) type reaction is fairly common; high energy is required to permit the proton to escape from the compound nucleus (i.e., intermediate nucleus formed after a neutron has been absorbed by the parent nuclei). In such reactions, e.g., $^{35}Cl(n, p)^{35}S$, the product nucleus has the same mass but its atomic number is one unit less than that of the target. The advantage of this type of reaction is that radioactive materials of high specific activity can be obtained because the product nucleus is different from parent one.

Another example of a high energy nuclear reaction is with fast neutrons (possessing energy about 10 MeV energy). Such neutron undergoes a $(n, 2n)$ type reaction, e.g.,

$$^{238}U_{92} + {}^{1}n_0 \rightarrow {}^{237}U_{92} + 2{}^{1}n_0$$

3.3.3 Nuclear Fission Reaction

Another important neutron-induced reaction is the fission of higher mass number elements, e.g., Uranium-235 and Plutonium-239, which is also accompanied by the emission of neutrons (approximately two or three neutrons per fission) and the liberation of energy. Consequently, it is possible for the process to be self-sustaining, and energy is generated continuously while the fuel is accessible. In the fission process, fissile material breaks up into approximately two equal parts, of mass numbers about 85 and 105, with a release of the total energy of about 200 MeV per fission. This energy is used for peaceful purposes, such as the generation of electricity. The assembly for generating energy for peaceful purposes is called a nuclear reactor.

3.3.3.1 Laboratory Neutron Source

In a laboratory, neutrons are produced either by (γ, n) or (α, n) reaction. In the (α, n) type neutron sources, an α-active radio-nuclei is thoroughly mixed with Beryllium and sealed in a robust-stainless steel container. Proper sealing is very important in order to make sure that radioactive emanation (e.g., radon from radium) does not leak out of the capsule. When in use, the capsule is shielded for γ-radiation and neutrons.

α-active nuclei used in neutron sources are Actinium-227 (such sources are cheaper), Radium-226, and Polonium-210. In each case, α-particles emitted from radioactive elements collide with beryllium, to give neutrons.

$$^{9}Be_4 + {}^{4}He_2 \rightarrow {}^{12}C_6 + {}^{1}n_0$$

The Ra/Be neutron source seems to be the most useful because α-particle density in the mixture is high, since various daughter products of Radium-226 are active. However, a Po/Be neutron source, which has the disadvantage of a fairly short half-life, is not associated with a lot of γ-radiation. A Pu/Be neutron source also has this advantage, coupled with a long half-life.

3.3.3.2 Interaction with Electromagnetic Radiations

Nuclear reactions initiated with electromagnetic radiations have been utilized to prepared laboratory neutron source. Antimony-124 is produced by thermal neutron (n, γ) by irradiation of Antimony-123. This radioactive isotope (Antimony-124) decays by γ-rays.

$$^{124}Sb_{51} \rightarrow {}^{124}Te_{52} + \gamma\text{-rays}$$

If this isotope is mixed with Beryllium-9, γ-rays emitted by Antimony-124 react to produce thermal neutron.

$$^{9}Be_4 \xrightarrow{(\gamma,n)} {}^{8}Be_4 + {}^{1}n_0$$

This source is made by the hot compacting of equal volumes of powdered Beryllium and Antimony, and the sources are sheathed in aluminum. They are activated by irradiation in a nuclear reactor to the required strength. An extra casing of the source with Beryllium increases thermal neutron density, and also decreases γ-radiation. Although this, in fact, is the cheapest neutron source available, has the disadvantage of short half-life (60.9 days), and high γ-flux, this means, after about 4–6 months, the ampule containing Beryllium and Antimony is radiated in the reactor to regenerate neutron activity. However, neutrons are slow, and require little moderating.

3.4 Particle Accelerators

Radioactive as well as stable isotopes are synthesized with the help of various types of accelerators. Some of the accelerators which are commonly used for carrying out high energy nuclear reaction are discussed here. For the synthesis of an isotope, first a suitable target material is selected like Carbon-12, and this material is bombarded by high energy nucleons (known as **projectiles**) like Helium atom or proton or any other heavier particles. The energy of these projectiles is increased with the help of accelerators that are is able to cross the nucleus barrier and enter the nucleus. Once it has entered the nucleus, a new isotope is created.

Normally, the projectiles are charged particles. These charged particles are kept in either a tubular or circular unit, which has the facility to create increasing potential. A gradual increase in the potential of an opposite charge forces the projectile to move with increasing velocity. Due to the attraction between the opposite charges, the projectiles gain acceleration and accordingly their kinetic energy increases. When the kinetic energy of the projectile is increased to a required value, it is allowed to bombard (i.e., interact with) the target material to produce the desired product. For gaining the required kinetic energy, the length of the tunnel through which projectiles move is kept very long. During the acceleration of the projectile, charges should not interact with the body of the long tunnel and lose their kinetic energy. To prevent the projectile from interacting with the body of the tunnel, a magnetic field is introduced along the tunnel. Magnetic field helps in making the projectiles move in a circular path without interacting with the main body of the tunnel.

There are different types of accelerators like Cyclotron, Linear-accelerator, Betatron, Synchrocyclotron, which are used for this purpose. Some of the common types are briefly discussed here:

Synchrotron is a cyclic particle accelerator. The magnetic field is time dependent and is synchronized to the particle beam of increasing kinetic energy. This was the first type of accelerator discovered which led to the development of various other types of accelerators. The largest accelerator is made in 2008 with a circumference of 27 km.

Linear accelerator (LINAC) is an instrument, which accelerates electrons by increasing their kinetic energy through a linear tube to a very high speed. This is used as a radiosurgery unit. The high energy X-rays generated in this system are used to destroy tumors.

Cyclotron is multistage particle accelerator. It operates under magnetic force on a moving charge. The magnetic field is altered to maintain the acceleration optimized. The movement of a particle is in a semicircular path. An electron is accelerated by the applied electric field.

Synchrocyclotron is a modified version of a cyclotron. The frequency of the driving radio frequency electric field starts as a cyclotron, which gradually increases till the desired energy is obtained. This helps to compensate for the relativistic effect when the velocity of the particles approaches the speed of light. Heavier particles like proton, deuteron, and α-particles can be accelerated by moving in a circle of increasing radii. This instrument can accelerate Deuteron to 200 MeV and even α-particles to 400 MeV.

3.5 Conservation of Mass and Energy

The energy released or absorbed in a nuclear reaction is calculated from the "concept of conservation of energy". As an example, the reaction $^{209}\text{Bi}(\alpha, 2n)^{211}\text{At}$ can be considered.

$$^{209}\text{Bi}_{83} \quad + \quad ^{4}\text{He}_2 \quad \longrightarrow \quad ^{211}\text{At}_{85} \quad + \quad 2^{1}n_0 \quad + Q$$
$$(208.980417 \quad (4.00260361 \quad (210.987496 \quad (2.01733088$$
$$\text{a.m.u.)} \quad \text{a.m.u.)} \quad \text{a.m.u.)} \quad \text{a.m.u.)}$$

Thus, the value of Q is equal to the mass difference between the two sides of the above equation, i.e., -0.021806 a.m.u., equivalent to -20.3 MeV. Since the value of Q is negative, some kinetic energy of the incident particle has been used to create products of greater mass than that of the target material. In other words, this reaction needs the incident particle to have energy at least as great as the amount absorbed in the reaction (i.e., 20.3 MeV). In such reaction, energy required for the bombarding particle is much less. For example,

$$^{232}\text{Th}_{90} \quad + \quad ^{1}n_0 \quad \longrightarrow \quad ^{233}\text{Th}_{90} \quad + \gamma + Q$$
$$(232.038211 \text{ a.m.u.)} \quad (1.008665 \text{ a.m.u.)} \quad (233.041428 \text{ a.m.u.)}$$

Here, the reactants are heavier than the product by an amount equivalent to 5.08 MeV. Thus, it is concluded that, in an exoergic reaction, incident particles need not be energetic. Thermal neutrons having energy of about 0.025 eV are sufficient to enable the neutron capture reaction to proceed.

3.6 Reaction Cross-Section

It is essential to calculate the time required for irradiation and decide the type of the target material suitable for getting a particular type of product in any of the nuclear reactions expressed earlier. In the foregoing discussion, we shall concentrate to understand some of the important parameters on which these nuclear reactions depend.

One of the important parameters is the probability of occurrence of particular reaction with the concerned target material. The probability of occurrence of a nuclear reaction is expressed either for the number of particles emitted, or nuclei undergone transformation, for a specific number of incident particles. A general method, which has been widely adopted to express relative efficiency for either of the processes by means of a quantity, is called **nuclear cross-section**. This represents a probability that a given nucleus will undergo a specific nuclear reaction. This probability term mathematically comes in a unit of cm^2 and hence it is called cross-sectional area for the reaction. The unit for cross-section is barn, and one **barn** is equal to 10^{-24} cm^2.

A knowledge of the magnitude of nuclear cross-section helps to calculate the amount of product formed by irradiating a target material with a nucleon. If for a particular nuclear reaction with an atom A, the nuclear cross-section is represented as "(σ)" to undergo a nuclear reaction to produce an atom B and if N is the number of atoms of A (irradiated), and "f" the number of incident particles per cm^2 per second, then the amount of product formed per second by the nuclear reaction would be $\sigma N f$, where f is assumed to be constant throughout the period of irradiation, provided the product is not a radioactive material. That is

$$\text{Rate of production of atom, } B = \sigma N f \tag{3.1}$$

But, if the product "B" is radioactive, it will decay with time as well, hence the net rate of production of "B" from "A" is given by

$$\frac{dN_B}{dt} = \sigma N f - \lambda_B N_B \tag{3.2}$$

in which N_B is the number of atoms of B present at time t, and λ_B the decay constant of B. On integration and simplification, Eq. (3.2), becomes

$$N_B = \frac{\sigma N f}{(1 - e^{-\lambda_B t})} \tag{3.3}$$

In this calculation, it is assumed that the atom "A" is not a radioactive material. However, if atom "A" is also a radioactive material which decays by a decay constant of A, then Eq. (3.3) will need to be modified. This equation thus can be used to calculate the quantity of radioactive material "B" formed after time "t" of irradiating material "A".

Many radioactive nuclei are produced by (n, γ) reaction in the nuclear reactor. In such cases, atom B is isotope mixed with atom A, and its specific activity is based on the unit weight of the target material. By interconversion of units, Eq. (3.3) can further be written in terms of activity of atom B produced:

$$A_B = \frac{0.602\omega\phi\sigma f}{3.7 \times 10^{10} A} \left(1 - e^{\frac{-0.693t}{t_{0.5}}} \right) \tag{3.4}$$

where

A_B = activity of nuclei B produced at time "t" in curies
ω = weight of target material
ϕ = isotopic abundance of nuclei which will produce the atom sought

for

A = atomic weight of target material
t = time of radiation, and
$t_{0.5}$ = half-life of isotope produced

A critical inspection of Eq. (3.4) can give a valuable information. In this equation, the term:

$$\frac{0.602\omega\phi\sigma f}{3.7 \times 10^{10} A} \tag{3.5}$$

is constant for a particular material and neutron flux. In other words, growth in specific activity will be of exponential nature as shown in Fig. 3.1.

If the time of radiation is long enough (about 5 to 6 times the half-life of the parent nuclei), the exponential term in Eq. (3.4) can be taken as approximately zero. The

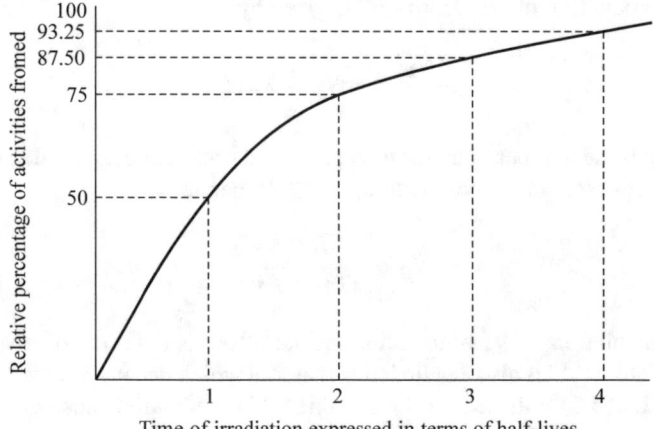

Fig. 3.1 The rate of approach to saturation in the production of a radioactive species as a function of the irradiation time in half-life units

specific activity of an isotope, thus, produced will be constant, and independent of time of irradiation. This suggests that maximum specific activity of the product can be obtained by irradiation (for a given amount of target material) for a period of 5–6 times the half-life of the expected radioactive isotope (which is to be produced by the nuclear reaction). It is also obvious from Eq. (3.4) that the activity of atom B can be increased by increasing weight "w" of the target. Alternatively, a high specific activity of atom B can be obtained by increasing the neutron flux. It must be remembered that with a long-lived isotope, it is not possible to irradiate material for a long time in the neutron source, because it will become very expensive to produce long-lived isotopes, e.g., for the isotope of half-life greater than few days or impossible for isotope with a half-life of a few years. With these types of materials, either neutron flux or amount of material or both are increased. Alternatively, for such cases, the Szilard–Chalmers process may be adopted to get high specific activity in a relatively shorter duration of irradiation. This aspect has been dealt with in a separate chapter.

If there are possibilities of the formation of more than one product during the nuclear reaction and the half-life of each radioactive isotopes is different, then the production of unwanted long-lived isotopes can be controlled by controlling the time of irradiation.

3.7 Some Features of Nuclear Reactions

In the study of nuclear reactions with any target material, it is advantageous to consider the following factors.

3.7.1 The Purity of the Target Material

If the target material is not isotopically pure, the products formed may be a mixture of the various nuclei (which in fact depends upon the cross-section and the isotopic abundance of each species), and the study of the reaction and the products formed becomes a mixture of products.

3.7.2 Conservation of Mass Relationship

The study of masses of reactants and the product involved in the nuclear reaction can sometimes help to rule out the possibility of the occurrence of certain reactions. For example, in the case of Bismuth-209 (mentioned earlier), the α-particle must have at least 20 MeV energy in order to give Astatine-211. In other words, Astatine-211 cannot be formed from Bismuth-209 by the nuclear reaction with helium ions of energy less than 20 MeV. A calculation based on the conservation of masses of

product and reactant can indicate the minimum energy of incident particle required
for carrying out the nuclear reaction.

3.7.3 Nature of the Products

It would be advantageous to know the characteristic properties of the expected nuclei
(such as half-life and types of energy of radiation emitted). This helps in planning
the manipulation of the products and the selection of counting methods.

The chemical properties are of assistance in selecting a suitable separation method
of the products from the target material. If a nuclear reaction is to be carried out with
a new element, the chemical properties can be predicted from its expected position
in the periodic table. This may be helpful in planning the experiment.

3.8 Applications of Nuclear Reactions

It is unfortunate that the advents of nuclear science first burst on the world in an
explosion over Hiroshima and Nagasaki. Nevertheless, the fundamental scientific
principles involved in the release of energy in nuclear weapons are now being widely
applied for peaceful purposes. The nuclear reactor is coming into use for the generation of power, such as electricity. Useful by-products from the reactor operation are
radioisotopes, produced either in the fission process itself or by the introduction of
other materials into the neutron atmosphere of the reactor. It is rather fortunate that
some of the radioactive isotopes, which have suitable properties for use as tracers
or radiation sources, such as Strontium-90, Cesium-137, and Barium-140, are produced in the fission process. Another important aspect of reactor operation is that
while the energy is produced for mankind during the nuclear fission reaction, a new
fissile material, Plutonium-239, is also produced. This material can also be used for
producing energy by the nuclear fission reaction.

$$^{239}\text{U}(n, \gamma)^{239}\text{U}$$

$$^{239}\text{U} \xrightarrow[23.5\,\text{min}]{\beta} {}^{239}\text{Np} \xrightarrow[2.33\,\text{days}]{\beta} {}^{239}\text{Pu}$$

Plutonium-239 makes a considerable contribution to the economics of the operation of nuclear power stations. Another fissile material is Uranium-233, which can
be produced from Thorium-232 by (n, γ) reaction in a nuclear reactor.

$$^{232}\text{Th}(n, \gamma)^{233}\text{Th}$$

$$^{233}\text{Th} \xrightarrow[24.2\,\text{min}]{\beta} {}^{233}\text{Pa} \xrightarrow[27\,\text{days}]{\beta} {}^{239}\text{U}$$

This reaction may become very useful, from the economic point of view, in a country like India, where Thorium occurs in plenty. The development of nuclear reactors based upon Thorium and Uranium-233 might be possible, thus avoiding the necessity to buy fissile and fertile materials from other countries.

There are numerous special applications of nuclear reactions in either analytical chemistry or even in the industry. Neutron activation analysis is one of these applications and is used to ascertain in a non-destructive experiment, the quantity and types of a minority element present in a sample. The method is to irradiate the sample with neutrons in order to induce radioactivity in the trace element present in the sample. The presence of the isotope is recognized by its half-life and radiation spectrum analysis. If a reference sample of a known weight has also been irradiated and treated in the same way, then the comparison of the two activities gives the percentage of the trace element present in the sample. An example is the determination of oxygen in steel by irradiation with 14.2 MeV neutrons from a neutron generator based upon a (n, p) reaction on oxygen.

$$^{16}\text{N} \xrightarrow[7.4\,\text{sec}]{^{16}\text{O}(n, p)^{16}} {}^{16}\text{O}$$

The short irradiation does not give time for appreciable activation of other elements, and the determination is carried out without destruction of the sample. The details of the activation process are discussed later.

Summary

This chapter has presented that nuclear reactions can be initiated by *(i)* charged particles (like doubly charged helium atom), *(ii)* uncharged particles (like neutron), and *(iii)* interaction with electromagnetic radiation. In order to carry out these nuclear reactions, we need neutrons and other interacting radiations. Special reference is made for generating neutrons in a laboratory. A brief description of some common accelerators used for carrying out nuclear reactions is also touched upon. The factors responsible to produce radioactive material and theoretical calculations to get an idea about the quantity of radioactive materials produced by these reactions are discussed. How the purity of a sample can affect the nuclear reaction, and some applications of nuclear reactions in analytical chemistry, are also featured in the chapter.

Chapter 4
Interaction of Radiation with Matter

4.1 Introduction

In the Chap. 2, we studied the nature of radiation normally emitted by radioactive isotopes (i.e., either α-particle, or β-particle, or γ-ray, or mixture of them). It is also seen that the number of such radiations emitted per unit time by an isotope is proportional to the number of atoms decaying per unit time. Therefore, for measuring the activity of a radioactive isotope, it is appropriate to design an instrument, which could detect and measure these radiations. Instruments, which are being used for this purpose, can measure either ion-pairs (i.e., pair of electron-positively charged ions) or visible light photons formed when these radiations interact with some suitable material. Therefore, it will be desirable to understand the effects caused by the interaction of radiation with a matter (i.e., the interaction of α-particle, β-particle, and γ-ray with a material) before discussing the principles on which different types of detecting instruments are based.

4.2 Types of Interactions

When radiation emitted by a radioactive isotope interacts with matter (i.e., the target), it is either scattered or absorbed by the target material. The transfer of energy from radiation to atoms of the absorbing material may occur by several processes, but among the commonly encountered ones, the following two are most important.

4.2.1   Ionization

This process results in the removal of an electron from an atom or molecule, thereby leaving the atom or molecule with a net positive charge. This causes the respective atom or molecule to become ionized.

4.2.2   Excitation

The process of excitation is the addition of insufficient energy (but sufficient enough to transfer an electron from the ground state to an excited state) to the target material to produce the ionization of material. The excited atom or molecule may lose its excess energy when the electron from its higher energy shell returns to its original ground state. When this occurs, the excess energy is liberated as photons which can escape or get absorbed in different parts of the target material. The nuclear excitation of the target material is of significance only for neutrons having relatively high energies and may not be of our interest.

4.3   Interactions with Particulate Radiation

A discussion on the types of nuclear radiations and their interaction can be conveniently divided into three main categories.

4.3.1   α-Particles

α-particles are ejected from atomic nuclei with a velocity in the order of 5% of light velocity. High mass number, charge, and velocity serve to make the α-particle an efficient projectile when it encounters atoms of an absorbing material and it has a high probability of interacting with the orbital electrons. Thus, the number of ion- pairs produced in unit length of track (i.e., the specific ionization) is high, and energy is rapidly transferred to the medium; its penetrating power is, therefore, comparatively poor.

α-particles may undergo either elastic (no transfer of energy) or inelastic (transfer of energy from the α-radiation to the target particle) collisions. Inelastic collisions result in ionization and/or excitation. After collision the kinetic energy of the α-particle is gradually dissipated by such interactions until eventually it captures two electrons and becomes a helium atom. Briefly, an α-particle is a highly ionizing and weakly penetrating radiation. α-particles from a given radioactive nuclei are emitted with the same energy; hence the range of α-particles, apart from straggling, will also be the same. The relationship between range and energy has been expressed empirically as follows:

$$\text{Range} = 0.318E^{\frac{3}{2}} \tag{4.1}$$

where range (expressed in cm) is the distance traveled by an α-particle in air at one atmospheric pressure and 15 °C, and E the initial energy of the α-particle in MeV.

4.3.2 β-Particles

β-particles can be either negatively charged or positively charged (known as positron). Positron is an electron with a positive charge; it falls in the category of particles along with electrons, as they have similar masses. Since they have equal masses and opposite charges, they lose their kinetic energy by a similar mechanism. However, the main difference is that a positron is annihilated with an electron of the interacting materials into an electromagnetic radiation of 0.511 MeV/particle.

Normally, a β-particle loses its energy in a large number of ionization and excitation events in a manner analogous to the α-particle. Owing to its small mass (and hence higher velocity at a given energy) and charge, there is a lower probability of a β-particle interacting in a given medium. Consequently, specific ionization is lower and the range at which the β-particle can penetrate into the interacting material is considerably greater than that of an α-particle of comparable energy.

Unlike α-particles, β-particles possess a continuous spectrum of energies, because decay energy of the nuclei is conserved between a neutrino and a β-particle. In other words, a β-emitter isotope emits β-particles of energies anywhere from zero to a maximum energy value, known as $E_{\max}$ (Fig. 2.1). $E_{\max}$ is the characteristic value for a particular radioactive nucleus. That is to say that a β-emitting radioactive isotope can be identified by measuring its $E_{\max}$ value because no two radioactive isotopes have the same $E_{\max}$ value.

4.4 Interaction with Electromagnetic Radiation

4.4.1 Electromagnetic Radiation

This group of radiations includes both X-rays and γ-rays. However, these two radiations differ only in their origin of formation and not in their mechanism of interaction. Here, only γ-radiation is discussed. γ-radiation is emitted from the nucleus with the velocity of light, has zero rest mass, and no electric charge. The specific ionization produced by this radiation is very small. The absorption of γ-rays in matter occurs by mechanisms which are completely different from the absorption of particulate

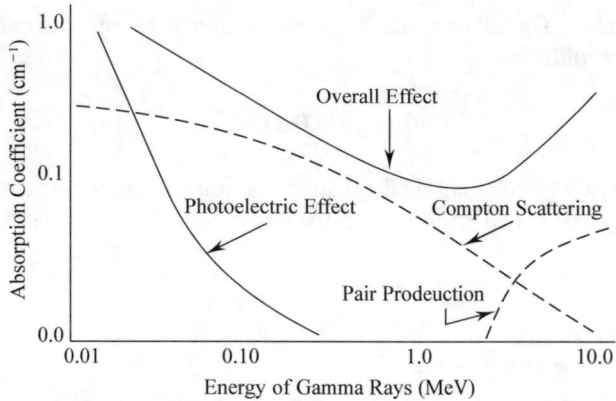

Fig. 4.1 Probability for γ-rays of various energies interacting with matter in different ways

radiations. γ-rays can lose whole or part of their energy in a single encounter. The absorption of a γ-ray by matter is exponential in nature and unlike a β-particle, there is no quantity corresponding to a range (E_{max}). The absorption of γ-rays in matter can occur by three different processes, depending on their energy, i.e., by Photoelectric effect, Compton effect, and Pair production.

4.4.2   Photoelectric Effect

The absorption of low energy γ-rays is mainly due to the photoelectric effect. In this interaction, the photon (i.e., γ-radiation) gives up all its energy to an atom, and an atomic electron is ejected. The kinetic energy of the electron is equal to the difference between energy of the incident photon (i.e., γ-radiation) and the binding energy of the electron in the atom from which it was ejected. The ejected electron transfers its energy to the medium in the fashion as described for β-particles. The probability that γ-rays (Fig. 4.1) will undergo photoelectric effect follows a specific relationship:

$$\frac{Z^5}{E_\alpha^{7/2}} \tag{4.2}$$

4.4.3   Compton Effect

For medium energy photons (energy > 0.501 and < 1 MeV), the most probable energy transfer mechanism is Compton scattering. A photon transfers a part of its energy to an orbital electron of the interacting material and this electron is ejected out of the orbital. The scattered photons leave the material with the remaining energy. These scattered photons may either undergo a series of similar Compton scattering (until their energy falls below 0.51 MeV) and eventually end up in a photoelectric-type interaction, or get scattered away into the space without having the chance to interact with interacting material. Thus, the probability of Compton scattering decreases with increasing energy of γ-rays (Fig. 4.1). The ejected electrons from the orbital of interacting material, on the other hand, transfer their energies in a fashion as described for a β-particle (refer Sect. 4.3.1).

If the spectrum of the energy of photons is measured, the Compton scattering would give a continuous "background" with one broad spectrum due to ejected electrons (of a type similar to β-spectrum) followed by one specific sharp intensity due to photoelectric effect (refer Fig. 2.3 in Chap. 2).

4.4.4 Pair Production

Photons of energy greater than 2×0.51 MeV (0.51 MeV is the energy equivalent to one electron) may be absorbed by pair production, where γ-photon of 1.02 MeV is converted in the vicinity of a nucleus, into a pair of electrons, one positive (known as positron) and one negative (like a β-particle). This process is known as **pair production**. If a γ-ray possesses energy greater than 1.02 MeV, the excess energy is shared equally between the two electrons as kinetic energy. This negative electron like a β-particle can now interact with the target, producing secondary ion-pairs. The process of the formation of secondary ion-pairs continues until the electron has lost all its kinetic energy. Likewise, the positron also produces ionization with the interacting material (*i.e.*, it can eject the orbital electron of the interacting material), until its kinetic energy has become zero.

However, after the positron loses all its kinetic energy and slows down to almost zero, it encounters a free electron of the interacting material. During the process of this interaction, the positron and electron of the target material get annihilated, converting their masses into energy. This energy appears in the form of two γ-rays, each possessing 0.51 MeV energy. These γ-rays may interact with the target by the process of Photoelectric effect. However, if by chance the newly formed two γ-rays possess energy higher than 0.51 MeV, they may interact with the target material by the process of Compton scattering as well. The probability of pair production increases with the energy of γ-rays (Fig. 4.1).

Hence, if the energy spectrum of γ-rays with energy greater than 1.02 MeV is measured, in addition to a peak corresponding to its actual energy, we would also observe low energy photons (appearing as background noise) due to the Compton scattering process, a pair production photopeak of energy 1.02 MeV (due to the annihilation of positrons with electrons), and a peak corresponding to the photoelectric effect (due to the interaction of photons of energy equal to 0.51 MeV). The resolution of all these peaks would depend upon the sensitivity of the measuring instrument.

4.5 Consequences of Interactions

From these discussions, it is obvious that when any radiation either electromagnetic (X-rays or γ-rays) or particulate (α-particles or β-particles) interacts with matter it can either ionize the material (due to the removal of electrons from the interacting material) or cause either a photoelectric effect, pair production or excite the material (due to insufficient energy transfer to the atom) to a higher energy. So, two processes can take place when such radiations interact with matter: ionization causing the formation of ion-pairs or the excitation of atoms of the interacting material.

4.5.1 Process of Excitation

When the interaction follows the process of excitation, the excited atoms of the material soon fall to its original ground state by emitting photons of equivalent energy (i.e., excess energy gained by the interacting radiation). These photons interact with matter by three processes discussed earlier. The creation of excited species by these radiations can easily be done with solid or liquid materials.

4.5.2 Process of Ionization

When the interaction follows a process of ionization, then depending upon the energy of the radiation, primary and/or secondary ion-pairs are formed. The kinetic energy associated with ion-pairs then becomes equivalent to the energy of the electromagnetic radiation. Since it would be difficult to measure the kinetic energy of ion-pairs formed in a solid, one normally uses gases as a constituent of target material for the interaction. This immediately imposes a limitation to detect γ-rays by this process, because the probability of their interaction with the gaseous particles is very low.

4.6 Types of Counters

For measuring the radioactivity of any isotope, we shall have to develop a counting instrument which could detect either the presence of ion-pairs or photon emitted by the excited species. The instruments which operate to measure either ion-pairs or photons emitted due to the interaction of radiation with matter are of two types:

1. Counters based on the production of ion-pairs in a gas, e.g., the **ionization chamber**, the **proportional counter**, and the **Geiger–Müller counter**. These counters are discussed in Chap. 5.
2. Counters based on the production of excited species which in turn produce Photoelectric, Compton effect and Pair production in suitable materials, e.g., **scintillation counters** which are discussed in Chap. 6. Under this class, we also have some special type of counters like semiconductor counters, which are discussed in Chap. 7.

Summary

In this chapter, we studied the type of interactions with matter which take place with various types of radiations (α-particles, β-particles, and γ-rays). Based on the properties of interactions with matter, counters for detecting these radiations have been touched upon.

Chapter 5
Ionization Counters

5.1 Introduction

Radiochemists handling radioactive materials must have a sound knowledge of various techniques used to detect and measure them. Furthermore, if a radioactive isotope is to be used as a tool in any chemical process, one should be able to identify the isotope, detect its presence and then analyze it quantitatively. In order to accomplish these requirements, radiochemist is required to select a suitable radioactive material, decide the form (liquid, solid or gas) in which the sample is to be measured for its activity, select a suitable counting system for the purpose and finally must be able to take a decision about the period for which the counting should be done to minimize the error of counting, etc. Therefore, in this chapter, discussions are made in a sequence, which would help the reader to understand the various aspects of ionization counting system and its limitations in counting some specific radioactive isotope. When radiations interact with surrounding media (as discussed in Chap. 4), they can produce ion-pairs, because they have enough energy to knock out electron from the outermost shell of the interacting atoms. Argon gas, for example, after the interaction with radiation produces a positively charged Ar^+ ion and an electron:

$$\text{Radiation} + Ar \longrightarrow Ar^+ + e^-$$

Number of Ar^+ and e^- (known as ion-pairs) thus produced is directly proportional to the number of interacting radiation. Therefore, if there is an instrument to measure number of ion-pairs formed, then the amount of radiation and hence the amount of radioactive material can be measured. Since the instrument counts the number of ion-pairs formed, the radiation must interact, with a specific volume of the enclosed gas. These conditions are met in an ionization counter.

© The Author(s), under exclusive license to Springer Nature Switzerland AG 2021 53
M. Sharon and M. Sharon, *Nuclear Chemistry*,
https://doi.org/10.1007/978-3-030-62018-9_5

5.2   Ionization Counter

Although the design of ionization counter can be of varied types, in general, it comprises two electrodes across which a DC potential difference is maintained. In between these two electrodes, a mixture of gas is enclosed which provides a sensitive volume of the detector. These electrodes are usually of plane, parallel or coaxial cylinders, or of multiwire geometry. Ionizing radiations, interacting with the gas, transfer energy mainly in the form of excitation or ionization of the gas. The ionization of gas produces ion-pairs (i.e., electron and positively charged ions). The respective electrodes of the counter attract these ion-pairs. Each ion-pair when deposited at the respective electrodes of the counter produces a dc current, which is measured to find out the number of radiations responsible to produce the ion-pairs. There are four types of ionization counters as mentioned below and each of them has their characteristic properties.

1. Ionization Counter
2. Proportional Counter
3. Geiger–Müller Counter
4. Spark Counter

5.3   General Design of an Ionization Chamber

Ionization chamber is made of stainless steel and its outer body is earthed. It comprises two electrodes. The body of the chamber acts as one electrode (normally earthed), and another one (normally acts as an anode) is enclosed in the chamber containing argon gas (Fig. 5.1). This central anode carries a positive potential insulated from outer casing by an insulator (y). At the end of the anode wire, there is a glass bead to stop any discharge of current between the anode and the window (x).

The window is made up of a thin aluminum sheet, which is in contact with the outer metallic casing of the chamber. The chamber is filled with argon gas at some suitable pressure (but not more than one atmospheric pressure). The anode is connected through a variable power source (V). Potential applied to the anode is measured across resistance (R).

5.3.1   Current–Voltage Characteristics of the Ionization Chamber

When a radioactive source is kept near the window, its radiation penetrates through the window into the active area of the chamber producing ion-pairs. These ion-pairs have a natural tendency to interact within themselves forming thereby a neutral argon gas. But in the presence of electrostatic field created between the two electrodes,

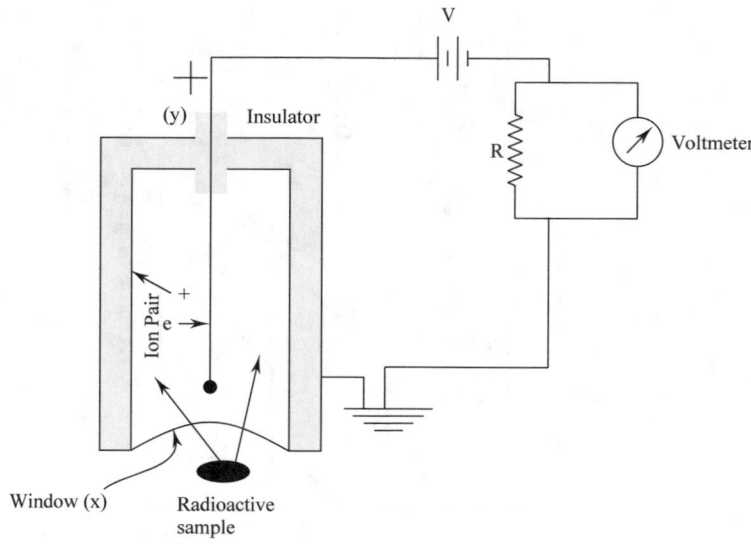

Fig. 5.1 Schematic representation of an ionization chamber

electrons and positively charged argon ions (produced due to ionization of argon gas) are attracted toward the anode and the cathode, respectively. These ions are thus deposited at the respective electrodes. During the deposition, a small current passes through the external circuit. Further increase of potential accelerates the ion-pairs so much so that they produce further ion-pairs by interacting with enclosed gas in the chamber. These ion-pairs are called **secondary ions pairs**. As the number of ion-pairs is increased with increase of potential, current increases accordingly. Finally, at some high potential, secondary ion-pairs are produced in such a large number that a continuous discharge takes place in the chamber. A typical current–voltage characteristic is shown in Fig. 5.2.

The entire current–voltage characteristic curve can thus be divided into six types. These six regions of curve are

1. Recombination region
2. Ionization region (In this region, ionization produced due to α- and β-particles can be differentiated.)
3. Proportional region (Like ionization region, ionizations produced due to α- and β-particles can be differentiated.)
4. Region of limited proportionality
5. Geiger region (Unlike ionization and proportional regions, ionization produced by α- or β-particles cannot be differentiated.)
6. Continuous spark region

We shall now discuss each of these six regions.

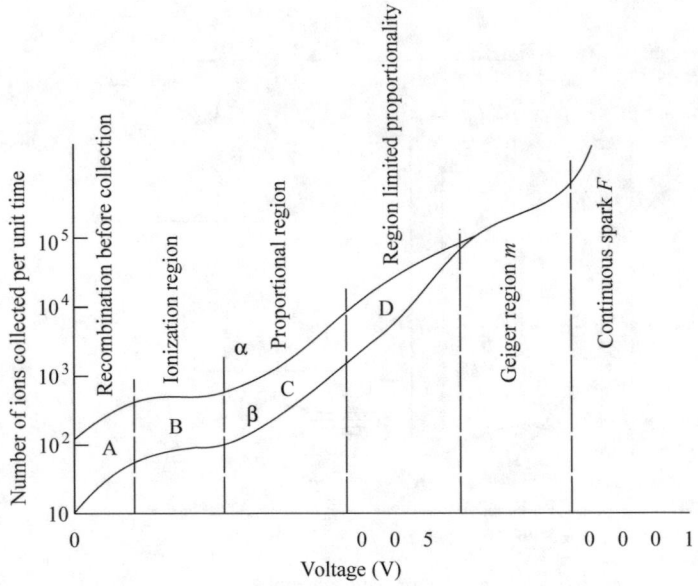

Fig. 5.2 Current–voltage characteristic of an ionization chamber. The upper graph is for α-particles and the lower for β-particles

5.3.1.1   Recombination Region

When a small voltage is applied to the anode, a small fraction of ion-pairs are collected. In this region, insufficient electrostatic field between the anode and the cathode is not able to prevent the ion-pairs from recombination before they have a chance to get collected at the respective electrodes. This region is called the recombination region (Fig. 5.2A).

5.3.1.2   Ionization Region

As voltage to the anode is increased, the electrostatic field increases permitting collection of more ion-pairs. It must be mentioned here that at this stage, production of ion-pairs does not depend upon the anode potential. However, percentage of these ions collected at the anode depends upon the anode potential. Thus, anodic current gradually increases with increase in the anode potential, until all ion-pairs formed by ionizing radiation are able to get collected at their respective electrodes. Any further increase in potential (about 100–200 volts) does not have much effect on the magnitude of the current. In other words, we can observe a very small plateau (*B*). The plateau *B* is called a _Saturation value_ or the region of the saturation voltage. Current at the plateau is due to ion-pairs (these ion-pairs are often called **Primary ions**) formed by the radiation. This region is also called **Ionization region**.

5.3.1.3   Proportional Region

When the voltage is further increased, primary ions are accelerated due to large electrostatic field developed across the two electrodes. These accelerated primary ions on interaction with gas molecules produce further ionization of gas molecules. These new ion-pairs are called **secondary ion-pairs**, and process is known as **secondary ionization** of gas molecules. It must be remembered here that number of ion-pairs formed by radiation has not altered. Only the number of ion-pairs was formed initially by radiation, due to increase in the electrostatic field that gets amplified. The process of multiplication of ion-pairs to produce secondary ion-pairs is also called gas **amplification**. As a result, the number of ions which reach the collecting electrode becomes more than initially produced by the interacting radiation. Consequently, current starts to increase with increase in the potential (C). If voltage is further increased, more and more secondary ion-pairs are produced. Current due to ions collected at the anode rises rapidly with increasing voltage. This increase in current collected at the electrode is proportional to the voltage. This voltage range is referred to as the **proportional region**. Like the ionization region, in this region also, ion-pairs produced per unit volume of gas is higher for α-particles than for β-particles. This is because the probability of interaction of α-particles with argon gas is more than β-particles due to heavier mass of the former particles. Because of this, for the same number of α-particles and β-particles, the former will produce more number of ion-pairs. Hence, number of ion-pairs formed per unit volume would be more with α-particles than with β-particles. Thus, in this region, one can differentiate ionization produced due to α-particles from β-particles. This region has one more added advantage over the ionization region; the magnitude of current is much higher in the proportional region as compared to ionization region. Hence, one does not need a sophisticated instrument to measure current in the proportional region.

5.3.1.4   Region of Limited Proportionality

Beyond the proportionality region, there is no strict proportionality between voltage and amount of charge collected. This region is called the **limited proportionality region**, and is shown in the segment (D). In this region, the slope of curves for the two types of radiation (i.e., α-particles and β-particles) are not the same. This is because the amount of charge that can be collected is limited by the particular characteristic of the chamber in use.

5.3.1.5   Geiger Region

If the voltage is further increased, two curves coincide. It is observed that the charge collected is not at all dependent on either the type of radiation or number of primary ions initially formed. In fact, it depends only on voltage applied to the electrode. In this region, field intensity around the center of the electrode is so high that any ion formed, primary or secondary, gets accelerated enough to cause a chain of ionization process. Because of this, an **avalanche of ions** is created in the tube. In other words, radiation of any type (i.e., irrespective of α-particles, β-particles, or γ-rays) producing even a single ion-pair in the chamber would be sufficient to cause collection of almost same amount of charge at the electrode as a particle giving rise to several ion-pairs. Its amount does not depend on the type of radiation or energy of the particle. In this region, charge collected at the electrode increases slowly with increase in voltage. This region is known as the **Geiger region** (E). This region, therefore, cannot differentiate ionizations produced by either α-particles or β-particles or γ-radiations.

5.3.1.6   Continuous Spark Region

Any further increase in potential beyond the Geiger region causes a continuous spark between the anode and the cathode. This region is called **Spark region** where generally, applied voltage is in excess of a few kV. Usually, high voltage pulses produced by the ionizing radiation with a very short rise time can be counted in this region. Because as soon as such pulses are applied to the counter, instantaneously current is produced, which is measured. Counters operating in this region are called **Spark chambers**. They can be made of various designs. This is truly a "do-it yourself" detection system to meet the specific requirements of the user.

It would seem that the process of secondary ion production which is caused by acceleration of ions toward the collecting electrode would continue to increase the charge indefinitely as voltage is increased higher and higher (i.e., beyond region F). But it is not so. On increasing the voltage ionization and discharge of ion-pairs tend to spread over a successively larger portion of the tube. This limits the magnitude of electrostatic field, which can be generated in the tube. Furthermore, since electrons are lighter than positive ions (i.e., Ar^+ ion), the former are swept out of the collecting gas very rapidly, leaving behind an atmosphere of positive ions (positive ion sheath) around the anode. These ions give rise to what is known as "**space charge effect**". This positive ion sheath surrounds the central electrode (i.e., the anode) reducing effective magnitude of electrostatic field intensity between the anode and the cathode of the chamber. These two factors limit the number of ions, which can be collected at the anode. Because of these two factors, the multiplication of secondary ionization cannot continue to increase indefinitely with increasing voltage.

5.4   Nature of Gas to be Used in Ionization Chamber

It is also important to realize that in principle, one could use any gas in the ionization counter, but it must meet the following requirements:

1. It should require least energy of ionization, i.e., the binding energy of outermost electron of the atom should be low. Argon requires about 15.7 eV for its ionization. In fact, the magnitude of the energy of ionization limits the energy of radiation which can be measured by ionization counter.
2. The gas must have large atomic size to offer greater chance of interaction with ionizing radiation.
3. The gas should be inert so that it does not react with a material of the counter.
4. It should be easily available, economical, otherwise the counter will become expensive.

Argon is preferred in ionization chamber because it meets almost all these requirements.

5.5 Regions Suitable for Counting Purposes

From these discussions, it is clear that for detection and measurement of radioactivity, ionization counter operating in (i) ionization region (ii) proportional region, and (iii) Geiger region would be useful. In the foregoing section, we shall now devote time to these three regions only.

A word of caution is perhaps needed here. It will be seen later in the section describing proportional and Geiger region that there exists a flat plateau in both, whereas no such plateaus are shown in Fig. 5.2. The reason being that current–voltage characteristic as shown in Fig. 5.2 is obtained with an ionization chamber over a very wide range of potentials. If plateaus as shown in Figs. 5.9 and 5.17 were to be shown in Fig. 5.2 also, then X-axis scale would need to be so large that the length of the potential axis would become too long to be plotted.

In Fig. 5.2 two graphs are shown, the upper one is for α-particles and the lower for β-particles. Graph of α-particle is shown above the β-particle, as the former has higher specific ionization than the latter. In other words, for radioactive sample containing α-particles and β-particles of similar activity and comparable energy, α-particle will produce larger number of ion-pairs per unit volume in the ionization chamber than β-particles. Moreover, α-particles always have specific energy (Fig. 5.3AB) hence magnitude of current would be very specific. On the other hand, β-particles possess energy anything from zero to a maximum value E_{max}. As a result, the amplitude of the current produced in the chamber would also have values from zero to a maximum. This shown in Fig. 5.3CD. Therefore, with β-particles, one observes a spectrum of pulses with varying heights (Fig. 5.3D, small triangular pulses).

It should be noted that the slope of the curve in the proportional region corresponds to the rate at which the collected charge changes with increasing voltage. This value is same for an α- and β-radiation. The study of current–voltage characteristics curve (Fig. 5.2) suggests that ionization counters could be designed to operate at ionization plateau (B), proportional region (C), Geiger region (E), and spark region (beyond F). The last two regions cannot differentiate between the type of radiations (e.g.,

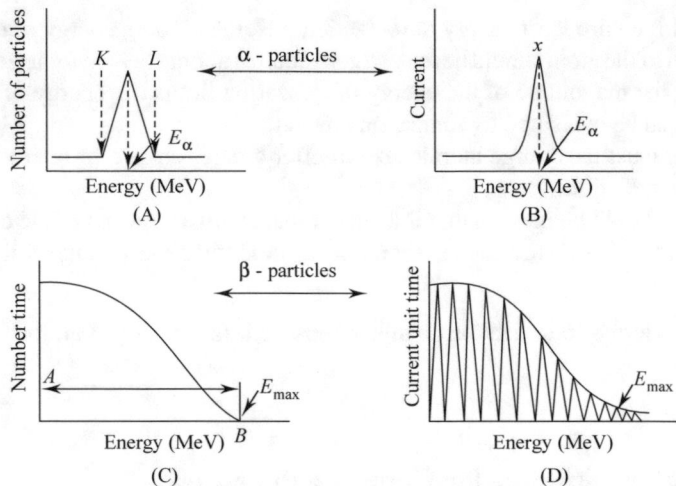

Fig. 5.3 Energy spectrum of α- and β-particles

α- or β-particle). The counter operating at either ionization plateau or proportional region can be used to distinguish α-radiations from β-radiations. The magnitude of current produced due to α- and β-particles are expected to be different for the reasons discussed earlier.

5.6   Nature of Pulses Produced in Ionization Chamber

Since the magnitude of current is represented by pulse height, and pulses produced by the α- and β-particles would be of different heights (Fig. 5.3), it may, therefore, be useful to spend some time to understand what we really mean by pulses and the pulse height before describing instrumentation techniques.

In electronic instruments, current is often measured as the number of pulses. For example, in houses, we have an AC power source with 220 V. We also say that the source has 50 cycles per second. What do we really mean by these terms in the language of electronics? A cycle (or a pulse) is represented in Fig. 5.3. The shape of current shown in Fig. 5.3A and D is related to the magnitude of energy of radiation producing ion-pairs in the ionization chamber. In electronic instruments, we normally measure number of potential pulses and their specific potential height. Hence, perhaps it may be a good idea to understand few characteristic properties of the current being measured in any electronic equipment. Some specific shapes of the current normally encountered in electronic equipments are represented in Fig. 5.4. Pulses no. I and II (Fig. 5.4) are called the **square pulses**. AB is called the **pulse height**. Its height is measured in terms of volt, and the time taken to rise the pulse from zero value (i.e., A) to its maximum potential "B" is called the **rise time**. This

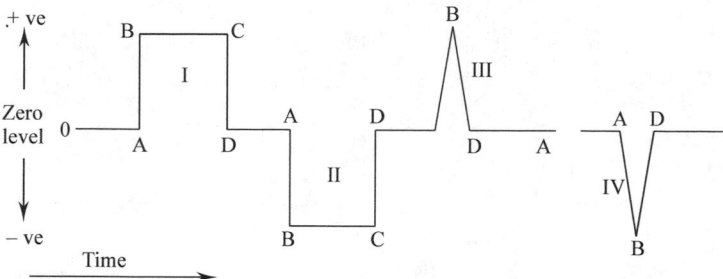

Fig. 5.4 Schematic representation of pulses produced in an electronic instrument. The shape of triangular pulses like III and IV are normally produced at output of the ionization chamber which needs to be converted into square pulses like I and II before they can be recorded by the instrument

height is also proportional to its energy. BC is the **duration of pulse** and is measured in the unit of second. This varies from a millionth of a second to a few hundredth of a microsecond or even seconds in very special cases. CD is the fall height. The time required to fall back from C to its original value D is known as the **fall time**. An ideal square pulse would have a rise time equal to fall time (i.e., $AB = CD$), and would be very parallel in nature. Number of pulses appearing per unit time is called the cycles and is represented by the unit **Hertz (Hz)**. The pulse could be in positive (I, III) or negative (II, IV) direction from its zero values. Pulses like III or IV have almost zero duration time but has its usual height (AB). Such pulses are difficult to be recorded by an electronic circuit, unless they are converted into a square pulse by allowing them to pass through a combination of resistance and capacitor of appropriate magnitude. When pulses are produced by ionizing radiation, they initially appear like pulses shown by III or IV. Energy of a pulse (also known as a **signal**) is designated by its height, because it represents the magnitude of the potential.

5.6.1 Conversion of Triangular Pulses to Square Type Pulses

If pulses are mixed with different energies, as is the case with β-particles, they are of varied heights (Fig. 5.5A). If all pulses are of the same energy, (as is the case of α-particles), then pulse height remains the same (Fig. 5.5B). For simplicity, we have shown square pulses in dotted line, as this type of pulse can be recognized by electronic instruments, and not by the original triangular type of pulses (as shown by full line). Therefore, by some suitable combination of resistance and capacitance (Fig. 5.5C, D), triangular pulses are converted to square pulses so that they do not lose their characteristic potential height. Moreover, in the previous discussions, we have seen that a current is produced when ionizing radiation enters the ionization counter. Since the electronic circuit can measure only the potential, the current has

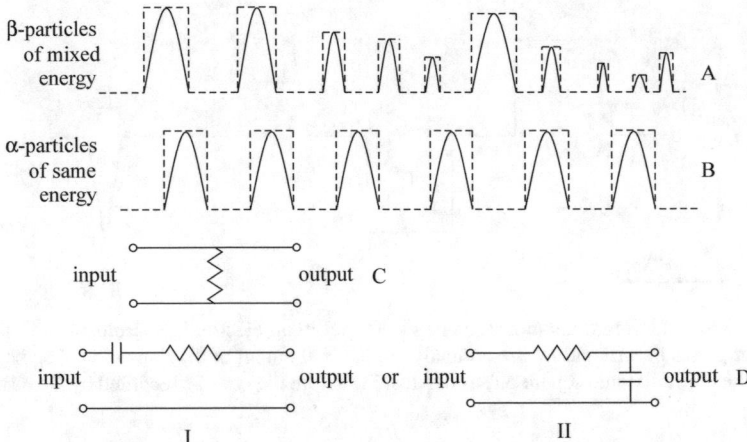

β-particles of mixed energy _____ A

α-particles of same energy _____ B

input ⎯⎯ output C

input ⎯| ⎯www⎯ output   or   input ⎯www⎯ output D

I   II

Fig. 5.5 Square and triangular pulses produced due to α- and β-particles: A—pulses due to β-particles, B—pulses due to α-particles, C—how current signal is converted to potential signal, and D—the arrangements of capacitance and resistance to convert triangular pulses into square pulses

to be converted into potential signal (pulse). This is achieved by following Ohm's law and using the circuit, somewhat like the one shown in Fig. 5.5C, D.

$$Current = \frac{Potential}{Resistance}$$

This denotes that in an electronic circuit (Fig. 5.5) current is converted into potential signal, whose shape need not be a square (normally of triangular shape). A square type signal from the triangular pulse is achieved by using either of the circuit shown in Fig. 5.5D (I or II), and by selecting a suitable value of a capacitance and a resistance such that it does not change the original pulse height. Accordingly, current obtained in an ionization counter is converted into a potential pulse. The magnitude of current (which is a measure of the energy of ionizing radiation) is calibrated in terms of height of the potential signal by allowing the current to pass through the circuit, as shown in Fig. 5.5C. The shape of this potential is converted to the square pulse by allowing the signal to pass through the circuit, as shown in Fig. 5.5D. It can be concluded that, height of a pulse is, thus a measure of the energy of the ionizing radiation and the number of pulses recorded per unit time is a measure of the intensity of ionizing radiation.

5.6.2 Pulses Due to α- and β-Particles

Depending upon the nature of radiation, the electronic instrument accumulates potential signals (square pulses) of various heights. Some instruments have the facilities to sort out the individual pulses (of specific height) and count their numbers. If α-particles of a specific radioactive material are measured, then the height of square pulses would be of the same height. If pulse height is calibrated in terms of MeV, then a measure of pulse height gives the information regarding energy of the α-particles. The number of such pulses recorded per unit time will give information about the number of α-particles recorded by the instrument per unit time (Fig. 5.5B). If these pulses are recorded on a strip chart recorder, a spectrum as shown in Fig. 5.3B will be observed. On the other hand, if β-particles are recorded, they will create square pulses of varied heights (Figs. 5.3D and 5.5A), and if instrument is connected to the strip chart recorder, a spectrum like Fig. 5.3C will be observed. The instrument, which can sort out these pulses according to their heights and keep an account of the number of pulses per unit time of each pulse height, is called **pulse height analyzer**. Cost of pulse height analyzer depends on the accuracy with which it can differentiate one pulse height from the other. If we can scale the pulse height in the range of 0–100, then the pulse height analyzer may divide this range into 100 channels (accuracy becomes 1 volt) or more (to make the accuracy for differentiating the pulses by less than one volt).

5.6.3   Relationship Between Energy of Radiation and Pulse Height

Based on the concept of pulse height analyzer, as discussed in the previous section and its application to current–voltage characteristic obtained with the ionization chamber (Fig. 5.2), the following conclusions can be drawn:

- The pulse height produced in an ionization chamber would be very small, in the order of micro to millivolts (up to B in Fig. 5.2). These pulses will have to be amplified to a pulse height of 1 volt before they can be registered by a suitable electronic circuit. Hence, a counter working in this region needs a sophisticated amplifier. However, counters operating up to this region can differentiate pulses produced by α- and β-particles.
- In the proportional counter, secondary ionization occurs; hence, pulse height initially formed due to primary ion-pairs are amplified to a large extent (region C of Fig. 5.2) than in an ionization chamber (i.e., region B). Therefore, though a proportional counter will also need an amplifier to amplify the pulses to 1 volt, a relatively less sophisticated amplifier is needed as pulse height is normally of millivolt height. However, counter working in this region will also be able to differentiate the pulses produced by α-particles and β-particles and pulses produced by other undesirable processes or background radiation. These latter pulses are called **noise pulses**.
- The pulses produced in Geiger region (region E) are amplified to such an extent that they lose their originality, and all pulses produced due to α-particles or β-particles become identical in pulse height (in the order of 1 volt). Therefore, a

counter operating in this region does not need any amplifier. This counter cannot be used to differentiate the pulses of α-particles from β-particles.

It may be appearing as a confusing statement that in ionization chamber the pulses are of microvolt range and in the proportional region they are of millivolt and yet we say that pulses produced by α- and β-particles can be distinguished in this region. On the other hand, pulses produced in Geiger region being of 1 volt cannot distinguish the nature of radiation. Fact is that pulses produced in ionization or proportional regions need to be amplified to 1-volt pulse height before they can be recorded by the instrument. What we mean by such statement is that the average height of pulses produced in the ionization chambers are of microvolt range, but within this range there are variations in their pulse height according to the energy of interacting radiation. When these pulses are amplified to 1-volt range, the variation in the pulse height is not lost. Same is the case with the proportional region. In the Geiger region, however, all pulses are produced by the interacting radiation to almost same pulse height and hence cannot be differentiated at all.

Now, we are in a position to discuss the three counters, as mentioned earlier, in somewhat greater depth.

5.7   Ionization Counters

Ionization counter operates in the region B as shown in Fig. 5.2. This counter can be made of various designs. One typical form of an ionization counter is shown in Fig. 5.6a. It consists of a metallic cylindrical chamber containing a central conducting electrode located at the axis of the chamber but insulated from the main body.

A proper voltage is maintained between the wall of the chamber (as a cathode, normally is earthed) and the central electrode (as an anode). The chamber is often filled with dry air at atmospheric pressure but other gases like argon may also be chosen for this purpose.

Since the ionization counter operates in the region B (Fig. 5.2), ion-pairs formed by the primary ionization would only be responsible to produce current. We have seen earlier, that pulses produced by these ion-pairs are of low magnitudes (of the order of microvolts) and hence a sophisticated amplifier is needed. Moreover, since the pulses produced by primary ionization process depends upon nature of the radiation (i.e., α-particles or β-particles, or γ-rays), they can be differentiated by the counter. For this purpose, a pulse height analyzer (or a discriminator bias, which allows the pulses to be recorded of height greater than the height set by the unit) is used, which distinguishes pulses of different energies. The basic problem in this counter is that pulses produced by radiations like α-particles or β-particles or γ-rays would be almost of same height as that produced by cosmic radiations (commonly known as the background radiation) present in the environment. A sophisticated pulse height analyzer must, therefore, be used to differentiate cosmic radiations from radiations

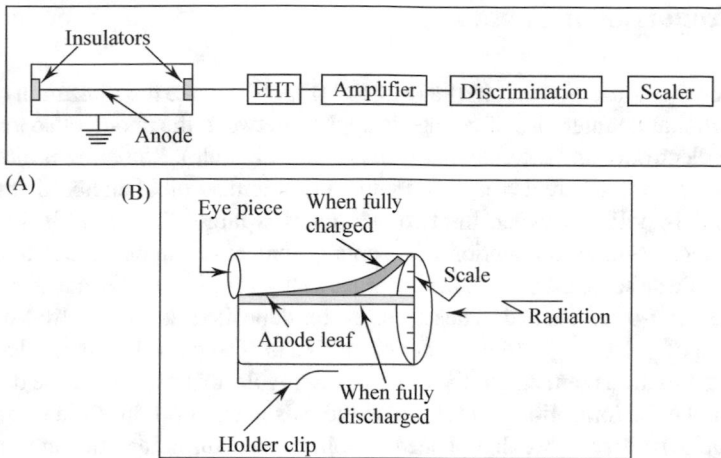

Fig. 5.6 a A schematic representation of a typical Ionization counter. **b** A schematic representation of a pocket ionization counter; used to monitor radiation received by the person carrying the counter

like α-particles or β-particles. This increases the cost of the entire setup of this counter.

We shall now discuss the process of conversion of ion-pairs into the pulse height. Electron produced in ion-pairs is collected at the anode of the counter. The magnitude by which the anode potential is lowered due to the deposition of electron is recorded as a negative pulse. This is amplified and with the help of pulse height analyzer pulse of different pulse heights are sorted out and finally recorded by the counter. This type of counter is used only for very specific applications.

A **pocket dosimeter** (Fig. 5.6b) which operates on the principle of ionization counter is used very commonly to monitor the dose received by a person working in the radiation area. This dosimeter instantaneously gives the magnitude of radiation received by the person. The chamber of pocket dosimeter has an anode wire, which is electrostatically charged in such a manner that the anode (needle type) points to zero of a scale. The scale is calibrated in rad/hr (unit of measurement of radiation, as defined latter). When radiation enters the chamber, primary ionization occurs, which leaks positive charge of the anode. This moves the pointer of the counter toward zero static charge. If total displacement of the pointer is calibrated in rad/hr, the measurement of displacement of pointer under a given condition shows the amount of radiation received in the chamber. Hence, a person wearing the pocket dosimeter can instantly find out the radiation received by him/her. This is a useful dosimeter because it instantaneously shows the dose received by the person present in the radiation area. The pocket dosimeter is designed to look like a pen, with a window at one end through which the displacement of pointer anode can be read (Fig. 5.6b).

5.8 Proportional Counter

Proportional counter operates in the region C (Fig. 5.2). Like the ionization counter, in proportional counter also, a voltage is applied between the anode (also known as collector electrode) and the chamber wall (i.e., the cathode). Radiation is allowed to enter the chamber, where it creates ion-pairs due to ionization of enclosed gas. Since the potential applied between the two electrodes is larger than what is applied in the ionization counter, production of secondary ionization cannot be ruled out. The number of these secondary ion-pairs becomes as high as 10^5 to 10^6 times the number of primary ion-pairs formed. These ions when deposited at the electrodes causes production of pulse height to the order of millivolts. The pulse height produced due to cosmic radiations remains of the order of microvolt (this aspect will be discussed later). Unlike the ionization counter, since the pulse height produced in this counter is in range of millivolts, we do not need a sophisticated amplifier, and hence the cost of this type of instrument is comparatively less.

5.8.1 Design of a Gas Flow Proportional Counter

Proportional counter can be designed in a variety of ways. Basically, there are two types of proportional counters, one with enclosed gas and hence has a window, and the another with continuous gas flow and without window. The construction of the former is very similar to Geiger counter (will be discussed later). The gas flow proportional counters have various geometries, usually not cylindrical. The anode is a fine tungsten wire, usually $0.001''$ in diameter. This wire is formed either into a loop or wielded to a terminal, which is supported by a single insulator or suspended as a straight wire between two insulators. The gas flowing through the counter is a mixture of $10:90$ methane : argon or $4:96$ isobutane : helium. However, in principle, other types of combination of inert gases can also be used provided the energy for removal of electron from their outermost orbital is lower than 30 eV. Description of a counter with a straight wire supported by two insulators is shown in Fig. 5.7. It consists of a half semi-spherical cylinder, with a hole in the center (Fig. 5.7). The sample holder has a thread through which the platform (Y) can be threaded into the counter, so that it is very near the central anode wire (L), or it can be threaded out up to a window point (Z), through which the radioactive sample to be counted can be loaded or unloaded from platform (Y). The size of the hole of window (Z) is such that when sample is near the central wire, the flat portion of one half semi-spherical cylinder is fully closed.

The ionizing gases are allowed to pass through (G) and get released into the atmosphere (if radioactive substance is non-volatile) through (H). In case of volatile radioactive materials, the gas has to be filtered by bubbling through a suitable solutions to remove radioactive material. It may be advisable, therefore, to connect the outlet (H) to a water bubbler. The out let of the bubbler is connected with a long

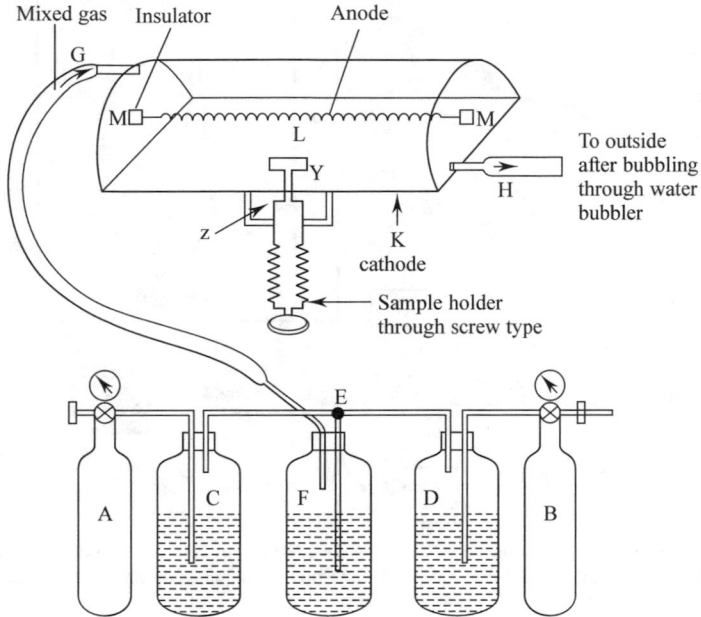

Fig. 5.7 A schematic representation of a gas flow proportional counter

tube so that gas escapes the counting room thereby preventing contamination of the laboratory and the personnel working therein. (A) and (B) are two gas cylinders, of argon and methane, respectively. Both cylinders through the gas regulator are connected with paraffin bubblers $(C, F, \text{and } D)$. If the height of paraffin liquid and size of the bubblers are same in each bubbler, then the ratio of $10:90$ of methane and argon can be maintained by adjusting the valves of the two cylinders such that 10 bubbles appear in methane bubbler and 90 in argon per unit time. The rate of flow of such mixture into the counter is adjusted by a pinch cork (E) which allows the gas mixture to enter the paraffin bubbler (F) at the rate of 1 bubble per second in the paraffin bubbler (F). The outlet of (F) is connected to the counter (G). Thus, with the help of gas regulator valves and pinch cork (E), the ratio of gas mixture and its flow rate into the counter can be adjusted. Body of the metallic cylinder (K) acts as a cathode. The central wire (L) which is insulated (M) from the main body of the counter acts as an anode, and is connected to other instruments as shown in Fig. 5.8.

5.8.2 Process of Ion-Pair Formation

When potential applied to the anode reaches a value where secondary ionization could occur, a single radiation which enters the chamber of the counter would produce large

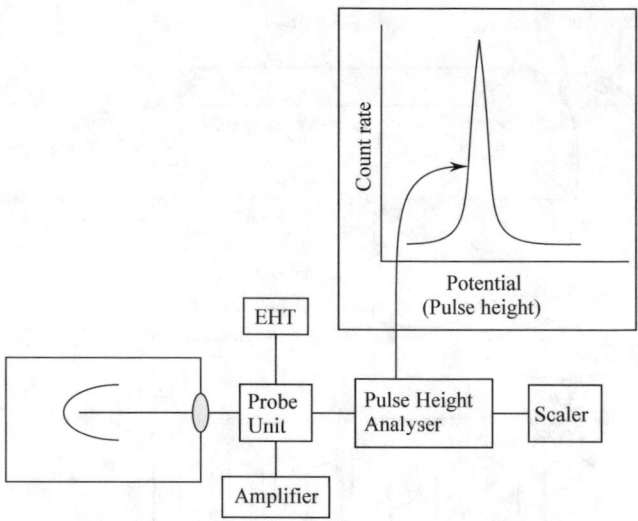

Fig. 5.8 A schematic block diagram showing arrangements of electronics instruments used for proportional counter. Inset shows the typical spectrum of α-particles obtained by use of pulse height analyser

number of electrons around the anode. These electrons, due to deposition of their charge at the anode, momentarily lowers the potential of the anode. The magnitude by which the potential is lowered appears as a negative pulse at the probe unit. An amplifier amplifies this negative pulse. The purpose of a probe unit is to maintain positive constant potential at the anode and separate the negative pulses produced by the ionizing radiation for their amplification. The probe unit, while separating pulses from the anode potential also, amplifies the negative pulses to some extent. The negative pulses are amplified and are connected either to a scalar unit for measuring the activity of the sample or to a pulse height analyzer, to sort out the pulses and their number in order to get the spectrum of radiation being measured.

5.8.3 Operating Condition

Before the counter can be used to measure the activity of a radioactive material, it is necessary to find out the most suitable potential, which should be applied to the anode. This is done by measuring the characteristic of the counter. For this purpose, a long-lived radioactive material is loaded into the counter (with the help of a sample holder Fig. 5.7 and count rate (number of negative pulses produced at the anode per unit time) is recorded with the scaler, (Fig. 5.8) and is measured as a function of applied potential to the anode (Fig. 5.9). If count rate is plotted versus the potential applied to the anode, in the beginning, as the potential increases, the count rate also

Fig. 5.9 A graph showing the variation in count rate with the applied potential to the anode of the counter

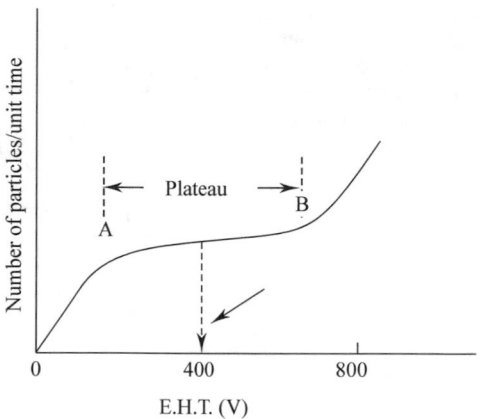

increases slowly up to some constant value (Fig. 5.9). Thereafter, the count rate remains almost constant even when the potential is increased by a value of about 100–200 V. Beyond this potential (Fig. 5.9B), the count rate starts to increase with increase in potential. The region where the count rate remains almost constant (i.e., region between A and B) is called the **plateau**. The middle of this plateau region is normally taken as the best potential for operating the counter.

If the spectrum of the radiation emitted by the radioactive isotope is to be determined, the anode potential is fixed to this value and then instead of reading the count rate by the scalar, a pulse height analyzer is used. With the help of this analyzer, we can measure the number of pulses produced as function of potential set in the pulse height analyzer. If a strip chart is connected with the pulse height analyzer, we get a plot of the spectrum of radiation directly (in set in Fig. 5.8).

5.8.4 Type of Radioactivity Measurable by the Counter

In principle, any radioactive material can be measured by this counter, but normally this counter is used to measure the activity of either α- or β-particle. γ-radiation is not preferably counted by this counter because it cannot produce as many ion-pairs as the particulate radiations. We would discuss this aspect again while discussing the Geiger counter. When the radioactive samples are mixed with α- and β-emitters, then the plateau for α-particles is observed at a lower potential while that for β-particles is observed at a higher potential (Fig. 5.10).

α-particles being heavier can transfer its energy to the ionizing gas much efficiently than the lighter β-particles. As a result, pulse heights of α-particles are larger than β-particles. Hence, secondary ion-pairs formed with α-particles can be recorded at much lower anode potential than β-particles. This is the reason for attaining plateau

Fig. 5.10 A typical graph showing current–voltage characteristic of a proportional counter for α-particles and β-particulate radiations

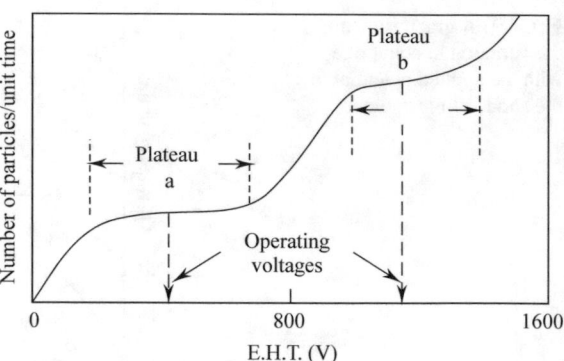

region for these types of radiations at two different anodic potentials. Moreover, slope of the plateau for β-particle is slightly higher than that of α-particle.

5.8.5 Background Counting

The background activity due to cosmic radiations can be found out by measuring activity in absence of any radioactive source. Since plateau for α-particles and β-particles lie at two different anodic potentials, the background activity must, therefore, be recorded at the corresponding plateau only. It is observed that at α-plateau, the background is of the order of 10–12 cpm (count per minute) compared to 30–40 cpm at β-plateau. The reason for this variation lies in the fact that at lower potential (i.e., at plateau of α-particles) number of ion-pairs formed by cosmic radiation would be much smaller compared to number of ion-pairs produced at plateau of β-particles. As a result, at operating potential of β-particles, ion-pairs due to cosmic radiation would be slightly more and hence a slightly larger background counts. However, the background can be reduced to even 4–5 cpm especially for α-particle counting with the help of a pulse height analyzer. Because the pulse height of pulse produced by cosmic radiation is much smaller than the pulse heights produced by α-particles. Another advantage of the proportional counter is that one can efficiently count the activity due to α-particles from a sample containing isotope emitting β-particles or γ-rays. This is achieved by using a pulse height analyzer.

5.8.6 Pulse Height Analyzer in Proportional Counter

Although we have discussed the pulse height analyzer, we shall discuss its application in some greater depth. The pulse height analyzer is a unit, which can be operated in two ways; it can separate pulses of individual height and help to count pulses of each height (known as **pulse height analyzer** or a **discriminator**) or it can give summation of all pulses (i.e., integral values of the pulses) above some preset pulse height values. In the first case, one can get an idea about the spectrum of energy, while in second one can get total activity by subtracting pulses from the set pulse height. The later form can be used to subtract background activity from total count rate of the radioactive material. In other words, under the integration mode it can separate all pulses above a certain height, whereas under discrimination setting it would allow the pulses to be grouped according to their respective heights.

For convenience, the pulse height analyzer has a range of 0–100 V, indicating that it has 100 channels and each channel represents a particular height of pulse, in terms of potential starting from 0 to 100. If each channel is calibrated in terms of energy, then one would get an idea about the energy of pulses coming out of each channel. There are pulse height analyzers, with 512 or 4000 channels, etc. Such analyzers can differentiate pulse heights of small energy difference. For example, pulses produced by an α-emitters of 2.4 and 2.35 MeV energies can be distinguished by 4000 channel analyzer more easily than with 100 channel analyzer. If pulse height analyzer is set at 10 volts, then it will allow all the pulses of height equal to 10 volts into the scalar. However, every instrument has its limit of tolerance, designated by a certain value of the window's width. If width of window is set to 1, then the limit of pulse height being allowed to pass to the scalar would be 10 ± 1.0 volt. Thus, width of the window helps in discriminating one value from another. Normally, the width of the window can be adjusted to 0.1, 1.0, or 10.0 as the case may be.

Now, we can discuss the application of pulse height analyzer in counting a radioactive sample either containing or emitting both α-particles and β-particles. β-particles have energy from zero to a maximum value called E_{max} (Fig. 5.3C), whereas α-particles have a definite fixed energy (Fig. 5.3B). Pulses produced by the β-particles will, therefore, be of different heights and their distribution would follow the same pattern as that of β-spectrum (Fig. 5.3D). Probability of interaction of β-particles with ionizing gas and hence, transferring its energy to the gas molecule in the counter is less as compared to α-particles of comparable energy. Therefore, height of pulses produced by β-particles would be lower than that produced by α-particles of comparable energy. Moreover, α-particles being monoenergetic, height of their pulses would also follow the same trend. γ-rays, on the other hand, (or cosmic radiation) though may have higher energy than any of these radiations, since they possess the highest penetrating power (i.e., probability of transferring energy to the ionizing gas will be less than even β-particles), their pulse height would be even smaller than that produced by β-particles.

A qualitative distribution of pulses produced by α-particles, β-particles, and cosmic radiation (i.e., the background radiation) is shown in Fig. 5.11. This will help to explain features of pulse height analyzer more clearly. Fig. 5.11 reveals the typical pulse height distribution produced by α-particles, β-particles, and cosmic radiations. The highest pulses of similar height are due to α-particles. The pulses produced by β-particles have a distribution of their height but are smaller than the pulses of α-particles. The background pulses are also of various heights but much smaller than pulses produced by α-particles. The energy of these pulses is expressed in volts (0–

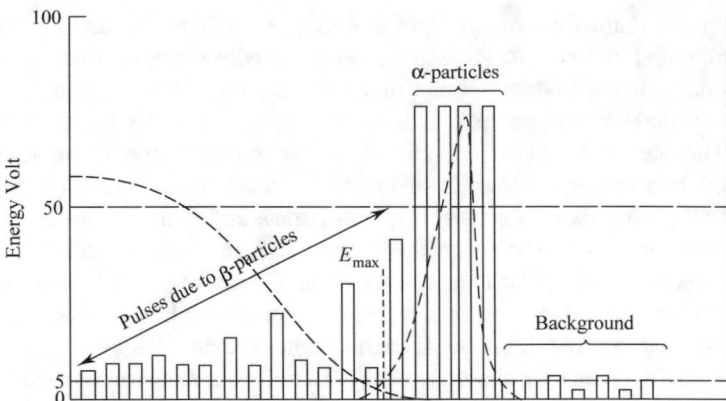

Fig. 5.11 A schematic representation of some hypothetical square pulses produced by α-particles, β-particles, and cosmic radiation (i.e., background radiations). Curve drawn with broken lines indicates the typical shape of a spectrum of α-particle and β-particles

100 V). From this typical graph, as shown in Fig. 5.11, the pulse height analyzer if used in an integral mode can eliminate the entire group of pulses produced by β-particles as well as those produced by background cosmic radiations. For example, setting the gate of the pulse height analyzer, around 50 V, would cut down all the pulses of height smaller than this value and allow the scalar unit to receive pulses of height greater than 50 V, which will be due to α-particles only. This process also discriminates the background radiation because their pulses would also be very small.

It should not be assumed that every time an α-particle is counted by the proportional counter, the pulse height would be greater than 50 volts. The pulse height depends upon the anodic potential and amplification factor used for counting purpose (Fig. 5.8). However, it is always true that under a given condition, pulse height produced by α-particles of an isotope would always be the same and would be much bigger than that of the β-particles (of comparable energy) or γ-radiation. This further adds a feather to the proportional counter, i.e., for counting α-particle one does not need any shielding of counter from the background radiations (i.e., cosmic radiation), because it can be eliminated by the pulse height analyzer.

5.8.7   Advantages of Gas Flow Proportional Counter

Although we have discussed some of the advantages of this counter, we enumerate them itemwise in greater depth for the benefit of the readers.

- Since gas is allowed to flow from one end of the counter to the other, the sample to be counted can be brought very near the anode wire, to minimize the losses of α-particles or β-particles due to self-absorption. This is because intensity of radiation decreases with the reciprocal of the square of distance. Hence, closer it is to the counter the greater is its chances of interacting with the ionizing gas.

- Due to the flow of gas in the counter, ions produced near the anode wire are constantly swept away, thus the anode remains practically free from the cloud of electrons to receive fresh ion pair produced by the radiation. This makes the dead time of the counter (this is related to the time for which the anode is covered with electrons of the ion-pairs produced by radiation) very small (of the order of 0.1 microsec.). The small dead time of counter permits the counting of high specific activity without much loss of activity. One could record with a proportional counter an activity of about 10^6 cpm, which is the upper limit of activity normally used in radiochemical works.
- Radioactive samples mixed with α-emitter and β-emitter can be counted without chemically separating the two isotopes if a pulse height analyzer is used.
- For counting of α-particles, no shielding of the counter from cosmic radiation is needed, whereas for β-particles, a small shielding by lead bricks is required.
- A gaseous sample of radioactive substance can be counted by the gas flow proportional counter. This is achieved by allowing the radioactive gas to flow with argon methane mixture slowly through the counter. For example, ^{14}C present in CO_2 as $^{14}CO_2$ in a sample, or ^3H present in CH_4 as C_3H_4, etc., can easily be detected by this counter. Because of this feature, the gas flow proportional counter can be connected with the gas–liquid chromatography instrument to detect any radioactive species present in the sample. In other words, instead of using a thermal detector or a flame ionization detector in the gas–liquid chromatography, one can use gas flow proportional counter as a detector for detecting the radioactive samples. It is important to realize that the conventional detector of gas–liquid chromatography can detect samples up to 10^{-6}M concentration, whereas gas flow proportional counter can detect samples up to 10^{-12}M concentration.
- Geometrical efficiency of counting (i.e., total solid angle that can be covered by the counter to count the activity of radiation) can be improved from 2π to 4π, by constructing a counter as shown in Fig. 5.12.

Typical spherical type of gas flow proportional counters for 2π and 4π geometrical efficiency, as shown in Fig. 5.12, are self-explanatory. In this type of counter, the radioactive sample is kept in one plane. For such counter, either hanging type anode or horizontal type anode are used. The source tray is made of a polythene sheet on which the source is placed. When this counter (which is equivalent to 2π geometry) is duplicated, as shown in Fig. 5.12, 4π geometry can be achieved. The source is kept on one side of the polythene sheet, such that upper side of the counter behaves like windowless counter while the bottom counter behaves like a counter with a window of thickness equivalent to thickness of the polythene sheet.

- A proportional counter can be used to calculate the absolute activity (activity with 100% counting efficiency). However, some corrections are needed to account for the following factors:

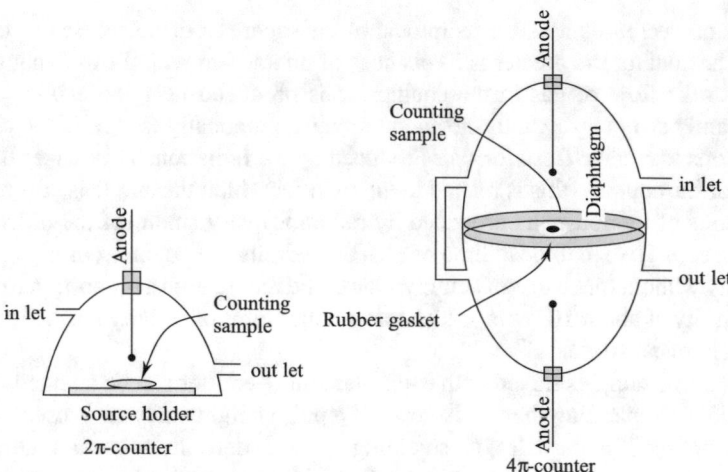

Fig. 5.12 Schematic diagram of 2π and 4π gas flow proportional counter

(a) **Geometry of counter:** Fraction of radiation is emitted from the sample that arrive at the counter.
(b) **Backscattering of radiations:** This causes an increase in intensity of radiation reaching the counter. Some radiation is reflected toward the counter due to backscattering from source holder.
(c) **Self-absorption:** Some intensity of radiation traveling toward the counter is lost while it passes through thickness of sample and air/ gas present in between the sample and the anode of the counter.

It is possible to evaluate the losses and gains in activity due to these factors.

- For the measurement of low energy β-particles, a proportional counter is capable of giving better resolution than a scintillation counter (see the chapter on scintillation counter). α-particles or β-particles of energy ranging from 250 to 100 KeV can also be counted with greater efficiency.
- The efficiency of counting α-particles is high. For β-particles of higher energy, efficiency very low (1–6%). This is because high energy β-particles produce less number of ion-pairs due to their high penetrating power. In other words, before the high energy β-particles have a chance to interact with gaseous molecules, it escapes from the active region of the counter. It is for this reason that γ-radiations are not counted by this counter. This is a good counter to measure activity of α-emitter or low energy β-emitter samples.
- A proportional counter can be made of any shape or size to suit the requirement. It can either be windowless, as described earlier, or with a thin mica window.
- It can be used to find the energy of α-particles, with the help of a calibrated discriminator unit (calibration of each volt of the channel is made in terms of eV energy).

5.9   Geiger–Müller Counter (G.M. Counter)

The G.M. counter is widely used because of its good characteristics, high sensitivity, and versatility for use with different types of radiation and energy ranges, large size of the output signal, and a reasonable cost. This counter has been built and operated successfully with tube having diameter of 1 ~15 cm, containing ionizing gas (2 mm to about 25 cm pressure of Hg) and with length from 1 to 50 cm. Nevertheless, here we shall be discussing the description of an End-Window G.M. counter, which is commonly used in radioactivity measurements. Since output pulses of the G.M. counter are typically in the order of a volt or more, at the most, one stage of amplification is needed before they can be counted. This makes the counter inexpensive and simple.

5.9.1   Design of the End-Window G.M. Counter

The G.M. counter tube, like proportional counter, consists of an envelope in which, two electrodes and the appropriate filling gas are incorporated. The internal collector electrode is a fine wire, which is a few hundredth of a mm in diameter. It is often made of tungsten because of its strength and uniformity in diameter. The anode (sometime called a collector) is usually a straight wire fastened from the insulators at both ends. However, in the End-Window G.M. counter, the collector is supported at one end only, while the free end is covered by a glass bead (which prevents point discharge of electricity to the window). Configuration of the G.M. tube is usually cylindrical, with the collector mounted coaxially. The other electrode, often referred to as the cathode, is generally a part of the envelope of the tube. If the envelope is metallic, it may serve directly as the cathode. If the envelope is of glass, its inside surface may be covered with a conductive coating to form a cathode. Stainless steel, nickel, or other materials are highly conductive and do not get oxidized quickly, hence makes a suitable cathode surface.

The most common filling gas is the noble gases, particularly helium, argon, and neon. Usually, a small percentage of additional gases are added for quenching purposes (this process is discussed in Sect. 5.13). One of the requirements for the satisfactory operation of a G.M. counter is that the electron attached to the enclosed gas should be easily detachable (i.e., the gas should be easily ionizable) so that only free electrons transfer negative charge to the anode. This is achieved by enclosing a gas, which needs very low energy to remove electrons from its outermost shell. The minimum energy required to ionize argon gas is 15.7 eV. This means that the radiations having energies greater than this can be counted by this counter. Unlike the gas flow proportional counter, with Geiger–Müller counter, the gas is permanently sealed in the anode chamber. Therefore, sealing of gas is done such that one end of the counter contains a very thin sheet of metal (usually aluminum) or mica sheet. This end is referred as the window of the counter. The efficiency of counting with

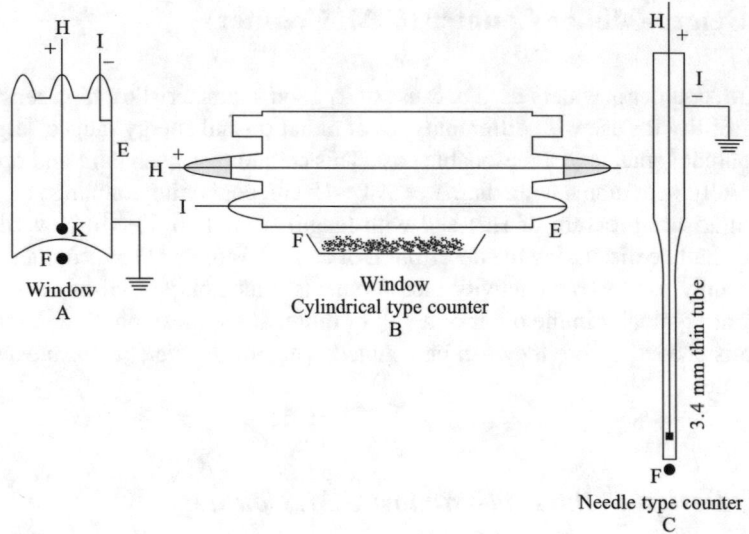

Fig. 5.13 Schematic diagrams of a typical End-Window counters of various shapes and sizes (A, B, C)

Geiger–Müller counter thus depends upon capability of the radiation to penetrate this window as well as its capability to ionize the enclosed gas.

The circuit diagram for Geiger–Müller counter is similar to that of a proportional counter, except that amplifier and pulse analyzer unit are not used, as pulses produced in this counter are of same height irrespective of the type of radiation being used. Some of the common types of Geiger–Müller counters are shown in Fig. 5.13.

5.9.2 Principle of a G.M. Counter

The principle of operation of a Geiger–Müller counter is exactly similar to the proportional counter, except that it operates in the region E (Fig. 5.2). The only difference is that the former counter cannot differentiate between the activities of α-particles, β-particles, or γ-rays. The counter can give a counting efficiency of 60–70% for the energetic β-particles (provided β-particles can enter the counter through the window), whereas only 1–4% for γ-rays. The dead time of this counter is normally 100–200 ms which puts a limit to the maximum activity which can be recorded by this counter (This aspect is dealt separately in foregoing sections). Because of these considerations, G.M. counters are preferred to count activities of energetic β-particles (preferably of energy greater than 0.5 MeV) and γ-rays only. Details of its operational part is discussed after explaining the description of the liquid G.M. counter.

5.10 Liquid Geiger–Müller Counter

Unlike the End-Window G.M. counter, a specially designed counter can be used to count liquid samples also. A typical design of a liquid G.M. counter along with a typical assembly of various electronics instruments are shown in Fig. 5.14.

The liquid G.M. counter consists of two coaxial glass cylinders. Space between the two glass walls is used for keeping radioactive liquid for counting. The inner glass cylinder contains ionizing gas, the anode and cathode of the counter. The cathode is either a spiral metal wire or the inner wall of this glass cylinder is coated with thin film of silver to act as a cathode. At the center of the inner glass cylinder, there is a

Fig. 5.14 A schematic block diagram of Geiger–Müller counter (end window and liquid type) with its various electronic units

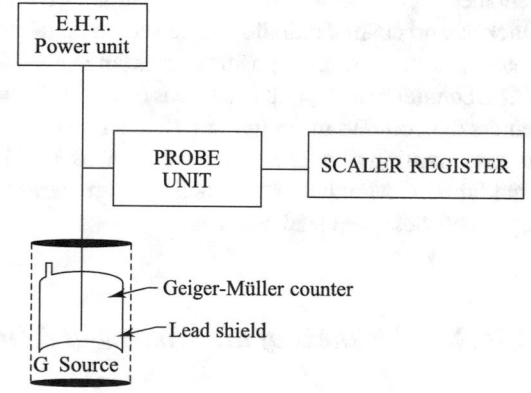

End-Window G.M. counter

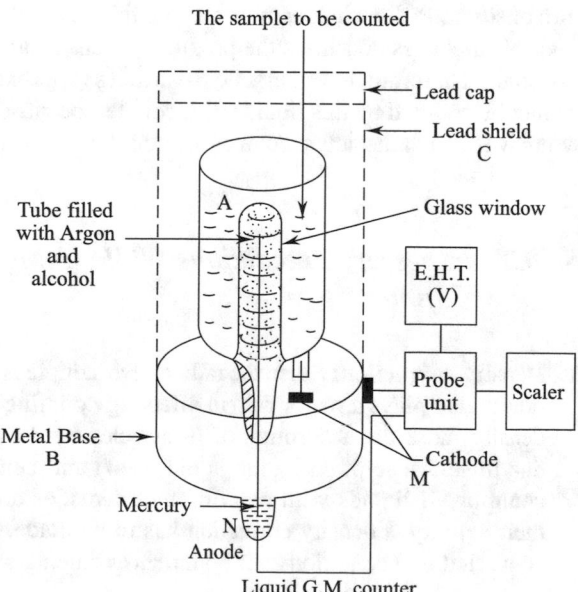

Liquid G.M. counter

coaxial anode wire. The inner glass tube, containing the cathode and anode is filled with a mixture of gases, argon and alcohol (Fig. 5.14). Usually, the space in between the two coaxial cylinders can accommodate about 12–14 ml liquid. The inner glass wall of the counter acts like a window in this counter.

5.10.1 The Assembly of the Liquid G.M. Counter

A liquid G.M. counter generally rests on a metal base plate (Fig. 5.13), such that the cathode (M) touches the metal base (B) and the central anode dips in mercury (N). Positive potential is applied through the mercury pool (N). The entire tube is shielded with lead bricks (shown by broken line, C), to reduce the activity due to background cosmic radiation. Care should be taken to see that no visible light enters into the counter when a positive potential is applied to the anode wire, as the liquid G.M. counter can respond to photons of visible light as well. It is observed that if light enters the counter, the count rate shoots up so high that the scalar gets jammed due to fast counting rate. This effect is known as **Joshi's effect**. Therefore, it is essential that liquid G.M. counter be protected from external light and background radiation by a well-designed lead chamber.

5.10.2 Thickness of the Window and Density Correction

It is worth noting that in a liquid G.M. counter, thickness of the inner glass and thin film of silver on the window makes a total thickness of the wall to about 50 mgcm^{-2} (unit of thickness considers the product of density and the actual thickness of the material). Therefore, low energetic β-particles (approximately lower than 0.5 MeV) cannot be counted by this liquid G.M. counter, because it cannot penetrate the glass window to reach the active zone of the counter.

5.10.3 Necessary Precautions While Using Liquid G.M. Counter

1. **Density correction:** Since the radioactive sample is in liquid form, the density of liquid also plays a major role in affecting counting efficiency. Higher the liquid density more the absorption of β-particles by the liquid (i.e., loss of radiation due to self-absorption by the liquid itself) and hence lower is the efficiency of counting. If liquids with two different densities are to be counted, to compare their activity, a density correction has to be made. Their activities recorded are converted to a condition as if both measurements were recorded in same type of

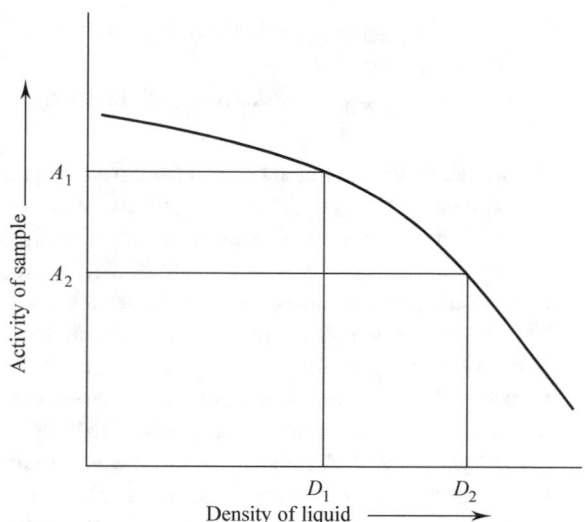

Fig. 5.15 A schematic graph showing the exponential variation in the activity of same quantity of long-lived radioactive isotope measured in liquids of different density, keeping the total amount of liquid same in a liquid G.M. counter

liquid. Such density correction is done in the following manner. Same quantity of a radioactive substance (either liquid or solid) is added to the different solvents of different densities, but the total volume of mixture is maintained to 10.0 ml, in each case. These solutions are counted in the liquid G.M. counter. Activity (i.e., count rate) is plotted against density of the liquid (Fig. 5.15). It is observed that the activity of the sample is not similar for all solutions, instead it decreases with increase in density of liquid.

The graph shown in Fig. 5.15 suggests that whenever the activity of a radioactive material present in more than two solutions are to be compared, one should either use a specific solvent so that there is no change in density or make necessary corrections in the activity by normalizing the count rate to a fixed density value. This type of correction is generally required in solvent extraction, where activities are distributed between two solvent phases (i.e., the activities in organic and water phases). In such type of experiments, it is important to make the density correction as discussed here.

A radioactive tracer of known specific activity is added equally into two liquids (of different densities) having equal volume. Activities of both are measured by a liquid G.M. counter. Let us assume that "A_1" is count rates, obtained with a radioactive material in 10 ml of a liquid of density D_1. Similarly, lets assume that we get a count rate of "A_2" of same radioactive material of same quantity added in 10 ml of another liquid of density D_2. Since known amount of activity was added to each solvent (A_0), the fractional loss of activities in each solvent are calculated as follows:

$$f_{\text{solvent } D_1} = \text{Activity loss in density } D_1 = \frac{(A_0 - A_1)}{A_0}$$

$$f_{\text{solvent } D_2} = \text{Activity loss in density } D_2 = \frac{(A_0 - A_2)}{A_0} \qquad (5.1)$$

These two factors can then be used for correcting the observed activity in the two experimental solvents. The actual activity recorded in each solvent should then be multiplied by the corresponding factor (i.e., $f_{\text{solvent } D_1}$ or $f_{\text{solvent } D_2}$ to get the actual count rate (i.e., count rate with density correction). These corrected count rates would be independent of the solvent density.

2. **Effect of volume of the radioactive solution**: The amount of liquid added to the liquid G.M. counter should always be the same. The liquid G.M. counter filled with liquid of different volumes gives different count rate, though each solution may contain same amount of radioactive material. This can be shown by the following experiment. A radioactive sample KI labeled with ^{131}I (1.0 ml) is added to a counter and its activity is recorded. Then, 1.0 ml of pure solvent is added to this counter, and activity is recorded after the counter is shaken to mix the solvent thoroughly. 1.0 ml of pure solvent is again added to this counter and corresponding activity is measured after shaking. This experiment is continued till no further solvent could be added to the counter (i.e., its inner tube is filled with the liquid). The volume of liquid which counter can accommodate is about 14.0 ml. The activity recorded for each volume of solvent is plotted against the volume of solvent, added to the counter. A typical graph showing the variation in count rate with volume of liquid is shown in Fig. 5.16.

   It is observed (Fig. 5.16) that after 8.0–10.0 ml of dilution, activity remains almost constant up to about 12–13 ml, but thereafter it shows a decrease in the count rate. The decrease in the activity after addition of 13.0 ml of solvent is due to poor geometry of counting. The radiations are lost due to self-absorption by solvent present above the central anode and also due to escape of radiations from the central compartment of the chamber, as the height of the solution is above the central anodic compartment (Fig. 5.14A). When the liquid is filled up to the upper part of the inner tube, a 2π-geometrical efficiency can be obtained, but when the liquid level is above the inner tube, radiations traveling in horizontal direction and upward direction escape the counter (i.e., away from the counter) hence are not counted. Due to these reasons, the counting rate decreases after a certain volume of liquid is added to the liquid G.M. counter. Hence, it is advisable to keep the total volume of liquid added to the counter always same, especially when activity of one sample is to be compared with the other.

3. **Washing of liquid G.M. counter:** It is necessary to emphasize that, in addition to washing with water, liquid G.M. counter should also be washed with solvent containing a non-radioactive isotope carrier. This process helps to decontaminate the counter completely from the radioactive isotope. For example, washing the counter with sodium bromide solution (0.01 M) when counting ^{82}Br isotope as bromide ions removes traces of radioactive ^{82}Br isotope adsorbed on the inner wall of the counter. The stable isotope must be in the same chemical state as that

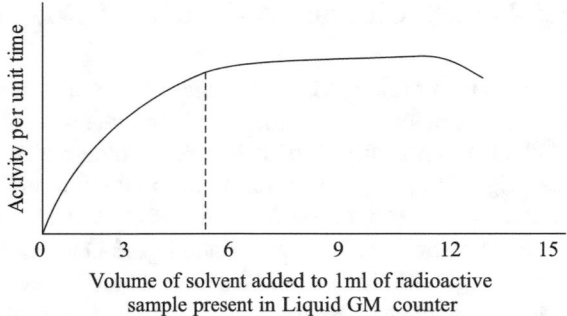

Fig. 5.16 A typical graph showing variation in the activity of 1.0 ml of KI labeled with ^{131}I versus addition of solvent into the liquid G.M. counter. After certain dilution activity decreases in spite of the fact that there is no change either in density of the solvent or the amount of the radioactive solution added to liquid G.M. counter

of the radioactive isotope, e.g., bromide ion will remove Bromide-82 ions and not Bromine-82 isotope.

4. **Amount of samples to be measured:** The amount of the radioactive sample should be such that when poured into the liquid G.M. counter, the upper surface of the liquid covers the top surface of the inner glass cylinder of the counter. This means that a small volume of a radioactive liquid cannot be counted efficiently, unless the radioactive liquid is diluted with a non-radioactive solvent to raise the level of the liquid up to the surface of the central cylinder (Fig. 5.14A). Moreover, it is advisable to maintain the volume of the solution to be counted, always the same (i.e., enough to fill the inner part of the counter), specially when large number of samples are to be counted and variation in count rate of sample are to be compared. The variation in the volume of the sample would change the contact area of the liquid with the glass wall of the counter, resulting in variation in count rate, in spite of the fact that all samples counted may be expected to have the same activity. Therefore, volume of the sample to be counted by the liquid G.M. counter should be such that it covers up to the central upper portion of the anode. However, sometimes volume to be counted may not be either same or not enough to cover the upper portion of the counter tube. In such cases, after adding the radioactive sample to the counter, a suitable solvent should be added into the liquid G.M. counter through a dropping micropipette, so that the liquid covers the upper portion of the inner tube. The advantage of pipetting solvent is that while the liquid is discharged to the counter through the nozzle of the pipette, due to the force with which liquid comes out of the pipette, the entire solution gets thoroughly stirred. The advantage of stirring the solution of liquid samples shall be dealt with subsequently.

5.11   Current–Voltage Characteristics of the G.M. Counter

Like the proportional counter for the Geiger–Müller counter also we need to determine the best operating potential. The current–voltage characteristic for Geiger–
Müller counter falls in the region (E) of the current–voltage characteristic of an
ionization counter (Fig. 5.2), but it appears from this graph that there is no plateau
in this region as observed with proportional counter (Fig. 5.9) which is not true. The
current–voltage characteristic of G.M. counter and liquid G.M. counter also gives
a plateau as shown in Fig. 5.17. The current–voltage characteristic for this counter
is obtained by adopting similar experiment as it was done with the proportional
counter. A radioactive sample (solid for the end-window and liquid for the liquid
counter) is kept near the window of the counter (or poured into the liquid counter)
and the arrangements are made as shown in Fig. 5.14 (for end-window counter and
for liquid G.M. counter). A potential is applied to the anode and activity is measured
in the unit of counts per minute. A plot is thus made of counts per minute versus
applied potential to the anode (Fig. 5.17). It should be remembered that half-life of
the radioactive isotope used for this purpose must be large (greater than few days)
so that there should be no change in activity of the sample while this experiment is
being carried out.

   Like the proportional counter a plateau is observed for this counter. The length of
plateau is about 200–300 V. Within the plateau region, count rate is almost independent of applied potential. The potential lying in the middle of this range is taken as
the operating potential for the counter. The slope of the plateau should not be greater
than 0.03 per volt. This slope is given by Eq. (5.2).

$$\text{Slope} = \frac{B - A}{V_2 - V_1} \times \frac{100}{0.5(A + B)} \tag{5.2}$$

where

   $A = $ count at the beginning of the potential (V_1) of the plateau.
   $B = $ count at the end of the potential (V_2) of the plateau.

Fig. 5.17   Current–voltage
characteristics of
Geiger–Müller counter
showing a typical plateau of
a good G.M. counter

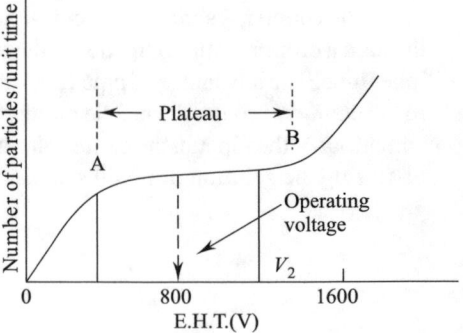

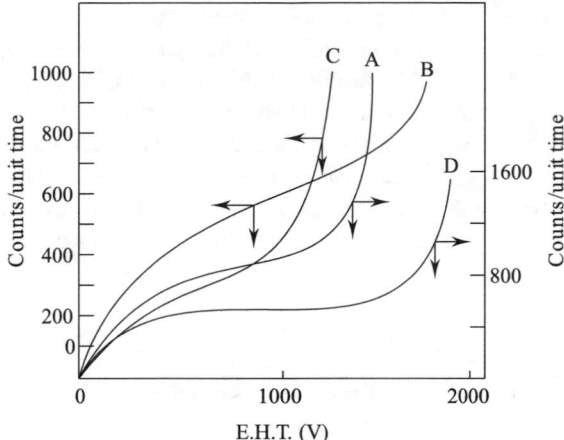

Fig. 5.18 Schematic representation of graphs showing variation in count rate with applied voltage to anode of a G.M. counter. *A*, *B*, and *C* show no plateau. When data of (A) is plotted on a suitable scale, it shows a plateau (D). Graphs (B) and (C) show no plateau even on scale used to plot (A), suggesting counter to be defective. It should be remembered that slope of plateau for graphs (A) and (D) will be almost the same and that of graphs B and C would be different

V_1 = potential at which the plateau starts.
V_2 = potential at which the plateau ends.

When we get the plateau as shown in Fig. 5.18, we have to take some precaution, before counting with this counter. A counter which is old and not functioning properly will show either no plateau (Fig. 5.18 (*B* and *C*)) or a very small plateau (Fig. 5.18A). A counter which gives a plateau like *B*, *C*, and *D* (Fig. 5.18) should be checked for its performance, and it is necessary to examine the following facts before arriving at any conclusion. In some cases, the counter may be operating properly, but due to faulty electronic connections it may behave erratically. Such erratic behavior can also give a bad plateau (*B*, *C* and *D* of Fig. 5.18). Whenever such plateaus are observed one should look for the following faults:

1. Electronic units like probe, EHT, and scaler units may be defective or loosely connected (if connected separately), because in some equipments, all these components are assembled in one unit.
2. The G.M. counter may be nearing its end of life.
3. In case of a liquid G.M. counter, the base at which cathode (Fig. 5.14M) or anode is connected through Hg pool (Fig. 5.14N) might be defective, or visible light might be entering the lead shield leading to some spurious counts (**Joshi effect**).
4. Scaling of the axis for plotting count rate versus potential might not be appropriate. For example, graphs *A* and *D* (Fig. 5.18) are plotted from the same data, but *D* shows a clear plateau whereas *A* does not. This behavior is related with the fact that the decay of radioactive material is a random process. Hence, there is bound to be a statistical fluctuation in the count rate. The scaling of *Y*-axis should be

such that statistical fluctuation gets minimized. If scaling of Y-axis is small, flat plateau may appear as shown in Fig. 5.18A though calculated slope may be 0.03 per volt. However, in order to observe visibly a plateau, list all count rates which appear almost same, take their average value, and then take the square root of the value. For plotting the data, make the scale of Y-axis such that 1 cm (or 1 inch as the case may be) represents a value almost equal to the square root of average count rate. If count rates are plotted under this condition, graph A can be made to look like D provided that the counter is free from other defects as mentioned earlier.

5.12   Dead Time of Geiger–Müller Counter

5.12.1   What is a Dead Time?

In previous discussions, we mentioned the term "Dead Time" and related it to the ability to measure the magnitude of activity of the radioactive sample. In this section, we shall take up this issue again and discuss its implications in a much greater depth. The concept of dead time is slightly abstract and may be difficult to understand, specially, if reader does not have much knowledge about the operation of electronic instruments. Therefore, an attempt to explain the concept of dead time is best by taking a simple example of a typewriter machine. When one key of a typewriter is pressed, a lever of alphabet, etc., is ejected giving an impression of the same on the paper. While learning typing, we often tend to press too many keys quickly and we know what happens then; the keys get jammed. Therefore, the typewriter stops functioning unless the keys are brought back to their original positions. We soon realize that a minimum time must lapse between pressings of two keys. This minimum duration required by the machine to operate is called the "**dead time**" of the typewriter. During this period, the machine cannot accept new commands. In other words, the total time required by the key to come back to its original position is the "dead time" of the machine. This dead time includes two factors: "operation time" (i.e., time taken by the lever to reach the paper soon after the key was pressed) and "recovery time" (i.e., time required by the lever to come back to its original position). Therefore, any machine working on a periodic mode will require a definite period to complete its one full cycle of operation before the next set of operation can start, and this total period is called the dead time of the machine. Likewise, after a Geiger–Müller or any ionization counter has received a radiation, it requires a definite period of rest before it can accept a new radiation to be recorded. The impact of this behaviour on operation of the counter is explained in forthcoming sections.

5.12.2 Impact of Dead Time on the Anode

When a radiation interacts with argon gas it produces ion-pairs which on multiplication due to several interactions, produces an avalanche of ion-pairs, in the form of
a cloud around the anode. The electrons being lighter than Ar^+ ions gets discharged
at the anode almost instantaneously. Collection of these electrons at the anode (positively charged) lowers the anode potential sharply by a value "x" (Fig. 5.19a). The
magnitude of "x" thus appears as a negative charge (referred as a negative pulse).
However, Ar^+ ions surrounding the anode take longer time to reach the cathode for
getting discharged. Due to this delay, anode potential takes almost the same time

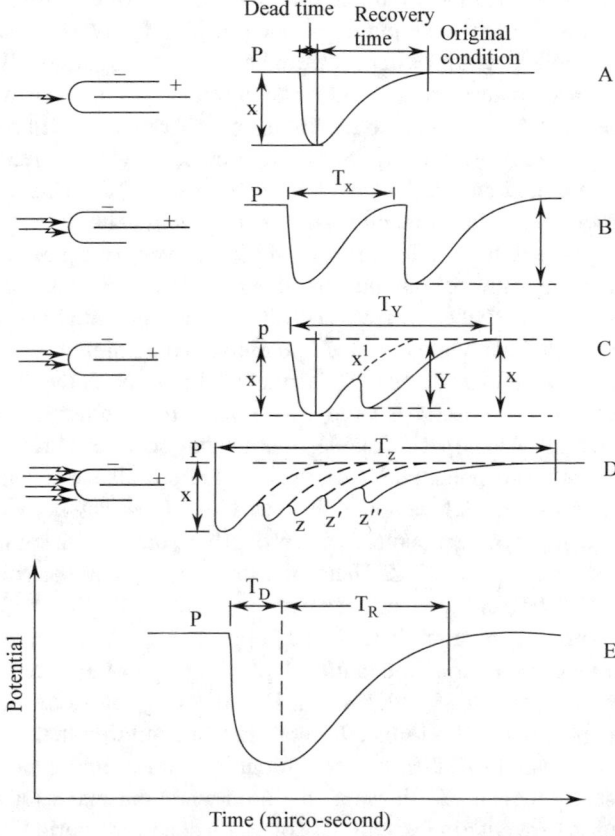

Fig. 5.19 A schematic representative of negative pulses formed at the anode of the counter, when
radiation of the radioactive sample enters the counting chamber. **a** shows the condition when the
counter receives one radiation causing a decrease in the anode potential and then its gradual decrease
in the magnitude of the negative pulse with time, **b** shows the condition when two radiations enter
the counter after a lapse of time t and **c, d** and **e** shows the impact of radiations entering the counter
much faster than time required by counter to regain its original position

to reach its original potential "P". Regaining of its original potential follows exponentially with time (Fig. 5.19a, b). Time taken by the anode to reach to its original potential (i.e., after electrons have been fully discharged at the anode) is related to the time taken by Ar^+ to get discharged completely at the cathode. After complete discharge of electrons and Ar^+ ions formed from a single radiation, the anode potential comes back to its original condition. At this stage if a second radiation enters the counter, the entire process, as discussed earlier, is repeated, and the output of negative pulses at the anode appears, as shown in Fig. 5.19b.

But, if a fresh radiation enters the counter, for example, at anode potential x^1 (i.e., much before the anode has regained its original potential), the radiation will experience a lower operating potential than what it would have experienced, had the anode reached its original potential "P". Consequently, number ion-pairs or cloud of charged avalanche formed too would be low. The number of electrons thus formed would also be less, thereby lowering the anode potential by a lower magnitude (y). In other words, anode potential will be lowered by a lower magnitude "y", than what a fresh radiation would have produced (x), had the anode potential been at "P". As a result, the time taken by the anode to regain its original potential will become larger T_y than T_x (Fig. 5.19c). Moreover, the negative pulse height "y" recorded by the second radiation would be smaller than the previous one, "x". This would mean a loss of count recorded in the scaler due to the anode potential being at "y" rather than at the operating potential "P". The situation becomes very complex when ionizing particles keep on entering the counting chamber one after another at such an interval that the anode gets no chance to recover to its original potential completely (Fig. 5.19d). That is to say that while the anode potential is regaining its original value, a third radiation causes it to go down to "z" from "y". In this case, the time required to recover the potential also becomes large (i.e., T_z) and the counter operates at much less anodic potential. The impact of anode potential has been seen in Fig. 5.17, where we saw that if this anode potential is low, the count rate is also low. This means that when the counter operates at lower potentials, we either loose activity while counting the radioactive sample or the counter is not properly operating during this period or the counter is dead for this period. Therefore, one can say that the counter is dead for a period of T_x or T_y or T_z, as the case may be.

A note of caution is required here. It may appear from the discussion that dead time of the counter is a variable quantity depending upon the specific activity of the radioactive sample. This is not true. In Fig. 5.19d, T_x or T_y or T_z is shown to explain the phenomena of dead time. T_z represents a condition that of a typewriter with many keys pressed together. Under T_z condition no counting can be done. T_x is the actual dead time, and T_y shows a condition when one starts losing the count because radiations are entering within the counter's dead time period. Dead time of the counter, therefore, can be divided into two parts (Fig. 5.19e).

$$\text{Dead time} = \text{Actual dead time } (T_D) + \text{Recovery time } (T_R)$$

The actual dead time (T_D) corresponds to a condition when the anode potential has been lowered down to a value "x", and the recovery time (T_R) corresponds to the time that anode would require to regain its original potential from the value "x".

5.12.3   Correction of Lost Counts

One would appreciate the impact of such dead time on counting a radioactive material if we take an example. A sample was recorded with a G.M. counter, and its activity was found to be 40,000 cpm. The counter had a dead time of 400 ms. We know by the definition, that after every one count recorded by the G.M. counter, the counter is dead for 400 ms. Therefore, while recording the counts of 40,000 cpm, counter would have been dead for $40,000 \times 400$ ms $= 16$ s. This means that 40,000 counts were actually observed only in $(60 - 16 = 44$ s) and not in 60 s. Therefore, actual number of counts recorded in 60 s should have been $(40,000 \times 60/44) = 54545.4$ cpm. In other words, 54545.4 cpm is the actual count rate recorded by the counter, if there was no dead time of 400 ms. Therefore, if dead time correction is not made, one would loose $(54545.4 - 40,000 = 15, 454.6$ cpm), i.e., a loss of activity by 28.3%. It is for this reason, that one should find out the dead time of the counter accurately and then correct the observed count rate by using the Eq. (5.3) as given here:

$$\text{Corrected cpm} = \frac{\text{Observed cpm} \times 60}{60 - \text{dead time in sec}} \tag{5.3}$$

5.12.4   Determination of Dead Time of the Counter

In proportional counter, the dead time is very short as compared to the G.M. counter because (i) the operating potential is much lower and (ii) the number of ion-pairs formed is also very low (for same activity of the sample). Both these factors force the anode to regain its original potential within a few microseconds, as compared to 100–400 ms in a G.M. counter.

It is important to realize that dead time of a G.M. counter depends upon geometry of the counter (size and relative distance between the cathode and anode, etc.). Hence, each G.M. counter has its own dead time. Therefore, it is necessary to find out the dead time of each counter. This is done by two methods. In one method, the dead time of the counter is determined experimentally every time the counter is used and the count rate is corrected for the loss of count due to the dead time. In the other method, the counter is connected to an external unit (usually the unit is connected between the probe unit and the scalar). This unit, after recording each count, lowers the anode potential by 300–400 V below the operating potential for a preset time, which is 3–4 order of magnitude more than the actual dead time. Because of this, the counter is externally made to become non-operational for a time set by this unit. This preset time is taken as the dead time of the counter, for calculation purposes. The latter method simplifies the counting procedure, otherwise, every time one uses the G.M. counter, its dead time has to be determined, which is a very tedious procedure as will be seen from the foregoing discussions.

1. **Measuring activities of two radioactive samples of almost same activity:** Two sources of long-lived radioactive isotopes with almost same activities are prepared for this purpose. They are made in such a way that the base of the source is a thin polythene sheet so that thickness of sources is almost negligible (therefore, absorption of radiation by the polythene sheet can be neglected). ^{32}P is taken as a source of measurement, but other sources like ^{89}Sr can also be used. The idea is that the source should be, as far as possible, a strong β-emitter. The activities of the two sources are measured separately by the G.M. counter whose dead time is to be determined. Let these activities be "m_1" and "m_2" for sources number one and two. Then activities of both sources are measured together. Let this activity be "m_{12}". Let us assume that "n_1", "n_2", and "n_{12}" are true activities of the source no. 1, 2 and when both were counted together respectively. Likewise, let "n_b" and "m_b" be the true and measured background activities, respectively. Since the background activities measured for all these measurements would be same, we can say that $m_1 = m_2$ and likewise $n_1 = n_2$. If τ be the dead time of the G.M. counter (in the unit of second), the time lost in counting various activities can be calculated as follows:

$$\text{Time lost in counting activity } (m_1) = m_1 \times \tau$$

The counter was thus operative for a period of $(1 - m_1 \times \tau)$ seconds during which m_1 activity was recorded. Thus, actual counts/sec., i.e.,

$$n_1 = \frac{m_1}{(1 - m_1 \times \tau)} \text{ per second} \tag{5.4}$$

Similarly,

$$n_2 = \frac{m_2}{(1 - m_2 \times \tau)} \text{ per second} \tag{5.5}$$

and

$$n_{12} = \frac{m_{12}}{(1 - m_{12} \times \tau)} \text{ per second} \tag{5.6}$$

and

$$n_b = \frac{m_b}{(1 - m_b \times \tau)} \text{ per second} \tag{5.7}$$

multiplying the numerator and the denominator of Eq. (5.4) by $(1 + m_1 \tau)$ and since numerically $m_{12} \tau^2$ is much smaller than unity (since τ is usually of the order of $(100 - 300$ ms$)$, one gets from Eq. (5.4)

$$n_1 = \frac{m_1}{(1 - m_1^2 \tau_2)} \approx m_1(1 + m_1 \tau) \text{ per second} \tag{5.8}$$

In order to evaluate this equation, we have to take the help of activities measured for both samples (i.e., m_{12}). n_b Eq. (5.7) can be assumed to be equal to m_b, because $m_b \times \tau$ would be a very small number and hence can be neglected. Moreover, $(n_1 + n_2)$ should be equal to $(n_{12} + n_b)$. Here, n_b is added because L.H.S. is counted twice (which would include the background twice). By substituting the values of n_1, n_{12} and n_b from the previous equations, we get

$$m_1(1 + m_1 \tau) + m_2(1 + m_2 \times \tau) = m_{12}(1 + m_{12} \times \tau) + m_b$$

or

$$\tau = \frac{m_{12} + m_b - m_1 - m_2}{m_1^2 + m_2^2 - m_{12}^2} \tag{5.9}$$

In this manner, the dead time of the counter (τ) can be determined. The main problem with this method is in making two sources having geometrically identical conditions with almost same activity and making sure that $m_1^2 \tau^2$, $m_2^2 \tau^2$, and $m_{12}^2 \tau^2$ are less than one. This derivation also clarifies how tedious it would be if we have to determine the dead time of the counter whenever we have to carry out the counting procedure.

2. **Electronic Quenching method:** The value of dead time determined by the above method is valid for the counter for which it was measured, because each counter has its own dead time depending on the geometry of construction and composition of gas filled in the counter. Moreover, dead time is not always constant; as the counter becomes older, this value changes slightly. Therefore, one will have to determine the dead time of the counter, prior to use. In radiochemical work, it will become very difficult to determine the dead time of every G.M. counter each time, before it is used for counting purpose. However, it is possible to have a quencher unit, which allows the anode to rest for a period, e.g., 400 ms or 600 ms, which is greater than the normal dead time of the counter. This means that after every count recorded, the counter will remain dead for a set of known period.

The set period is always kept greater than the natural dead time of the counter. The quencher unit after recording each radiation lowers the anode potential much below its normal operating voltage so that the counter cannot respond to the incoming radiation. After a lapse of say 400 ms, or any such time set by the quencher unit, the anode potential is automatically brought back to its original potential. Thus, counter always receives a new radiation at its operating anode potential. Observed count is then corrected according to the artificial dead time set by the quencher unit. The advantage of this method is that one does not have to determine the real dead time of the counter by a cumbersome procedure, as described earlier. Nowadays most of the scaler units are supplied with a quencher unit and its dead time can be set to a desired values like 400, 600, and 1000 ms. All observed counts are corrected for this artificial set dead time using Eq. (5.3).

5.13 Chemically Quenched G.M. Counter

In G.M. counter, positive ions (formed due to interaction of argon gas with radiation) migrate all the way from its point of formation to the cathode or from the vicinity of anode to the cathode. Due to potential difference between anode and cathode, these ions are accelerated and acquire some kinetic energy. Upon reaching the cathode, in addition to combining with electrons of the metal, they may also undergo collision with the cathode metal ejecting an electron per collision. These ejected electrons are also deposited at the anode, giving an additional count, which is recorded by the scaler.

In order to minimize the formation of such cathodic electrons, the kinetic energy of positively charged moving ions is decreased by allowing them to interact with some gases, which either are decomposed or dissociated into atomic gases, in the counter. This is achieved by enclosing some chemical into the counter, which preferentially undergoes inelastic scattering with positively charged ions approaching the cathode. Such collision reduces the kinetic energy of Ar^+ ions and thus prevents ejection of electrons from the cathode. Two types of G.M. quenched counters are commercially available: one filled with organic materials and the other with halogen gas.

5.13.1 Organic Gas Quenched G.M. Counter

Typical organic gases used for this purpose are ethyl alcohol (at 10 mm of Hg), or ethyl formate. These gases are known as **quencher**. The organic molecules being neutral and bulkier than charged Ar^+ ions provide larger area for inelastic collisions. These collisions break the organic molecules into small fragments. Kinetic energy of argon gas decreases to prevent ejection of electrons from the cathode. The disadvantage of using such gases is that they are gradually consumed (or destroyed) in the counter. The life of such counters, therefore, depends upon the concentration of organic gas enclosed. Such counters are found to count to about 10^9 and are known as **organic quenched G.M. counter**. It has been customary to make such type of counters with window made of aluminum. Such counters operate at an anode potential of around 1200 V.

5.13.2 Halogen Quenched G.M. Counter

Halogens like chlorine or bromine gases have also been used for this purpose. The main advantage of these gases is that, in addition to provide large area for collision, the halogen molecules after collision with charged ions produce halogen atoms, which again recombine to give the original halogen molecule. Because of this reversible process, unlike organic molecules, halogen gases are not consumed. This gives a very long life to the counter. In order to make an efficient inelastic collision with Ar^+ ion, bromine being the bulkiest molecule among the halogen gases, is preferred in these counters. Normally, this counter operates at 600–700 V. The only disadvantage with halogen quenched G.M. counter is that after some time halogen starts corroding the metal of the chamber. It is for this reason mica is used as window rather than aluminum.

5.14 Window Thickness

The end-window G.M. counters are made of either aluminum or mica windows, whereas liquid G.M. counter has a thin glass wall, 50 mg cm^{-2} thick. Therefore, thickness of these windows puts a limit on the energy and type of radiations which can be counted with these counters. As far as γ-rays are concerned, because of their high penetrating power, they are unable to produce ion-pairs efficiently in the counting chamber. α-particles, on the other hand, cannot penetrate the windows of the counter, hence they too cannot be counted efficiently, unless some modifications are made to the window. Hence, strictly speaking we are left with β-particles only, which can hopefully be counted with reasonable efficiency. In order to appreciate the effect of the window thickness, percentage transmission of β-particles of different isotopes through the window of different thickness are shown in Table 5.1.

Table 5.1 Variation in % transmission of radiations into the chamber of counter, emitted by various β-emitters through the various thicknesses of window of G.M. counters

Source	Max. Energy	Window Thickness (mg/cm^2)%		
β-particle/MeV		Transmission values are given here		
		0.9	4.0	30
C-14	0.154	79	35	20
Ca-45	0.250	89	61	2
Cu-64	0.580	97	87	35
P-32	1.717	99	97	79

The fifth column represents aluminum while the third and fourth columns represent mica windows of different thickness. It is obvious from Table 5.1 that for counting weak energetic β-emitters, mica window with a thickness about 0.9 mgcm^{-2} should be preferred. The unit of thickness is given in mgcm^{-2} because absorption of radiation depends upon thickness (cm) and the density of the material, i.e., mg/cm^3 $\times$ cm = mg cm^{-2}.

The liquid sample of weak β-emitter, however, cannot be counted by a liquid G.M. counter, because its wall thickness is around 50 mgcm^{-2}. It should be noted that the values given in Table 5.1 give the probability of transmission of β-particles through these windows, and it should not be confused with the efficiency of counting because efficiency of counting depends upon geometry of counting, source thickness, efficiency of ion-pair production inside the counting chamber, etc.

5.15   Limitations of Ionization Counters

Certain limitations of the ionization counters are enumerated here:

1. γ-rays being highly penetrating can easily enter the counter. At the same time, it can also escape the active area of the counter before it gets a chance to interact with the enclosed gas molecules to produce primary ion-pairs. Therefore, the efficiency for γ-ray counting with ionization counters is about 1% or less, whereas with β-particles or α-particles (provided they can enter the counter region), the efficiency can be as high as 80%.
2. Liquid samples of weak β-emitters or α-emitters cannot be counted by the liquid G.M. counter with a high efficiency because the wall thickness of the glass window is around 50 mgcm^{-2}. Windowless proportional counter could be used for such counting, provided a provision is made to load the liquid sample into the counter. Such arrangements are difficult to achieve due to greater chances of contaminating counting chamber during loading or unloading of sample. Therefore, there is a need to develop a counter specially to measure the activity of solid/or liquid samples of α-emitter, weak β-emitter and γ-ray emitters. This is discussed in the next chapter.

Summary

In this chapter, we studied the impact of particulate and electromagnetic radiation on gas under various potentials applied to a counter consisting of gas and two electrodes (anode and cathode) enclosed in a closed cylinder. How this behavior differs with α-particulate radiation and β-particulate radiation are also discussed. Based on these types of behavior, different types of ionizations counters have been developed (ionization counter, proportional counter, and G.M. counter). The principles of each of these counters and characteristic properties of each counter have been discussed. These ionization counters suffer from a factor known as dead time. How corrections are made for the count lost due to dead time of the counter has been discussed. G.M. counter can be used for solid as well as liquid sample for counting.

Chapter 6
Scintillation Counter

6.1 Introduction

From Chap. 5, we realized that although ionization counters are handy instruments to measure activity of radioactive samples, but it cannot be very useful for counting γ-rays or low energy β-particles or α-particles, because these types of radiations are incapable to either generate ion-pairs within the ionization chamber or penetrate the window of the counter to produce ion-pairs. Therefore, there is a need to develop counters to measure the activities of these types of radiations also. This is achieved by using a scintillation counter which is discussed in detail in this chapter.

6.2 Scintillation Counter

A scintillation counting system is a combination of electronic devices, by which flashes of visible light produced by the interaction of radiation with a fluorescent material are converted to voltage pulses of size which can be recorded by a scaler unit. The visible photons liberate electrons at the photocathode of a photomultiplier tube, which are multiplied by a series of dynodes, each held at potential more positive than the previous one. The multiplied electrons, thus, form a voltage pulse at the anode tube. The negative pulses (like the negative anode pulses generated in ionization counter) are amplified and discriminated, and are finally counted either by an electronic scaler unit or scanned and converted to a spectrum. A schematic diagram of the scintillation counter is shown in Fig. 6.1.

© The Author(s), under exclusive license to Springer Nature Switzerland AG 2021
M. Sharon and M. Sharon, *Nuclear Chemistry*,
https://doi.org/10.1007/978-3-030-62018-9_6

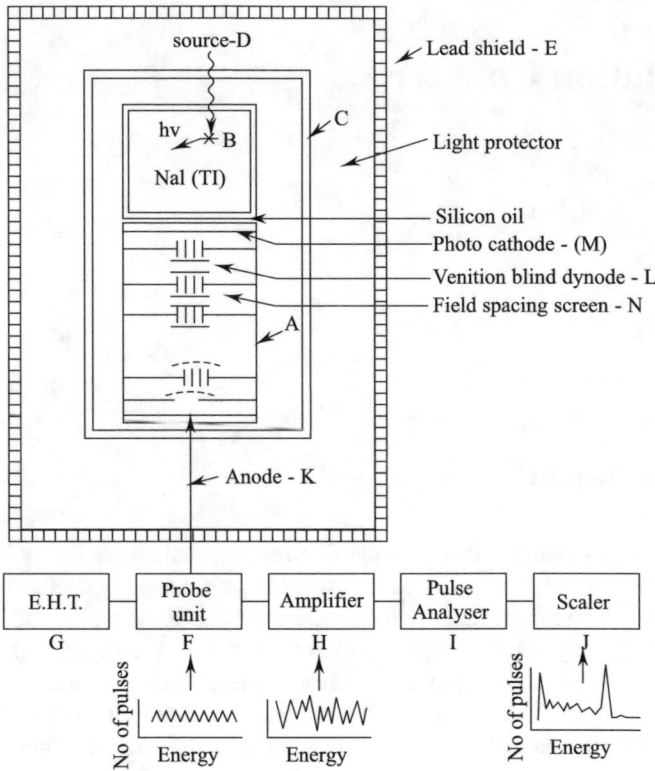

Fig. 6.1 Schematic representation of a typical NaI(Tl) Scintillation counter unit

6.3   Principle of Scintillation Counter

The scintillation counter was discovered much before ionization counters, but due to difficulties in measuring photons produced by the scintillation process, this counter did not become popular. Nevertheless, after the discovery of electronics instruments and techniques to count large number of photons per unit time, produced during the production of fluorescence by the interaction of radiations with phosphor materials, scintillation counter started to gain its importance in radioactivity measurements. Unlike the ionization counter, in scintillation counter, visible photons are generated by interaction of radiation with a suitable material called **phosphor**. These visible photons are measured with the help of electronics instruments. Phosphor (**fluorescent material**) on interaction with radiation produces excited atoms. These excited atoms *via* a series of processes transfer their energies from one state to another. Finally, a part of energy is converted into visible light, while remainder is dissipated as heat or by non-radiative processes. The output of visible photons depends upon energy transferred from the radiation to phosphor. Hence, like ion-pairs (formed in ioniza-tion counter), the number of photons and their wavelengths are proportional to the number of radiations and energy of radiation which interact with phosphor, respectively. These processes of excitation and generation of photons are completed in few milli-microseconds. Hence, unlike the ionization counter, there is no appreciable dead time for the scintillation counter. The photons are then converted into negative pulses, which later measured with the help of electronics devices, as described for the proportional counter. The basic difference between the scintillation counter and ionization counter is that the former relies upon production of photons from phosphor and the latter depends upon formation of ion-pairs from the enclosed gas in the counter. Both the photons or ion-pairs are then counted by converting them into electrical signals.

6.4 Components of a Scintillation Counter

The basic components of a scintillation counter are

1. A phosphor, which can produce visible photons per radiation by interacting with it.
2. A photocathode, which converts photons into electrons.
3. A dynode, which produces approximately 2.5 electrons per electron interacting with one dynode.
4. An electronic system which amplifies these electrons into an electrical pulse so that the pulse height becomes proportional to the energy and its population becomes equivalent to the intensity of the interacting radiation.
5. Amplification of these pulses and sorting them in similar fashion, as in the case of proportional counter.
6. Finally recording these pulses.

A block diagram depicting the setup of a scintillation counter is shown in Fig. 6.1. The system has a solid NaI(Tl) scintillator (B) (different scintillators can also be used) kept over a photomultiplier tube (A). The contact between the scintillator phosphor (B) and the photomultiplier base (A) is ensured by adding transparent silicon oil which is visible to visible radiation. A bad contact or air gap may reflect the visible photons away from the scintillator, preventing its interaction with the photocathode (M), and leading to a loss in count. The photomultiplier tube is protected from external visible light by either covering it and the phosphor with an aluminum cover (C) or wrapping it and the scintillator with a black tape. Under no circumstances should the photomultiplier tube be exposed to light when potential is applied to the photomultiplier (i.e., when extra high tension (E.H.T.) voltage is applied to it). This spoils the tube due to production of very high current. This is because the photocathode or dynode is sensitive to visible light and it cannot differentiate between light coming from the external source or phosphor.

The photomultiplier tube is also protected from external cosmic radiations by covering the assembly (C) with a lead brick house, having a wall thickness of 2–3″ (E), because a scintillator cannot differentiate between cosmic radiations and the radiation coming from the sample. The photomultiplier tube is connected to a EHT unit (G) through a probe unit (F). The output of the photomultiplier tube is connected through a probe unit to an amplifier (H), a pulse height analyzer (I), and a scaler (J).

6.5   Design of Photomultiplier Tube

There are different designs available for the photomultiplier tube. The design of a tube given in Fig. 6.1 consists of an anode (K), ten dynodes (L), and a photocathode (M). Between the two dynodes a field spacing screen (N) is used. The dynodes have also varied designs. In Fig. 6.1, dynodes of a venetian blind type is shown. The design and potential of the dynodes are made such that electrons can never be reflected in an upward direction. The photocathode is usually made of cesium and antimony oxide, as these materials can eject electrons when photon interacts with them. The entire assembly, i.e., photocathode, dynode, and field spacing screen are enclosed in a brown transparent glass and sealed under vacuum.

Each dynode is separately connected to a probe unit through which a high tension voltage is applied in such a manner that the dynode next to the photocathode has the lowest potential and the potential of other dynodes gradually increases till the last dynode (next to the anode) has the highest potential. Normally, a photomultiplier tube has about 11 dynodes.

The photocathode ejects approximately 2.5 electrons per photon incident on it (i.e., when one photon reaches its plate). In this way, one photon produces approximately two thousand electrons and these are deposited at the anode, which produces a negative potential pulse, similar to ionization counters. This pulse is amplified and analyzed using a pulse height analyzer. Energy of electrons thus collected at the anode and their number is proportional to the energy and the number of photons produced by the interaction of the radiation with the scintillator, respectively. The energy of photon corresponds to the energy of interacting radiation, while the population of electrons of a particular energy corresponds to the number of interacting radiation. Thus, by measuring the height and its corresponding number of negative pulses, we get an idea about the energy and intensity of the interacting radiation.

In brief, the operation of scintillation counter depends on the following consecutive events:

1. Absorption of nuclear radiation from a radioactive sample (D) occurs with the scintillator (NaI in this case) which results in excitation of the scintillator.
2. The de-excitation or ion re-combination produces a visible photon per radiation absorbed by the scintillator. Photons of wavelength equivalent to the energy of radiation are emitted. The magnitude of wavelength and its number depends upon the energy and number of radiations which produced excitation per unit time by the processes of luminescence.

3. Visible light photons are converted into electrons by an electronic eye, a photo-cathode (M), of a photomultiplier tube (A). Absorption of a light photon by the photocathode produces photoelectron of corresponding energy and intensity.
4. The electron-multiplication process by a venetian blind type dynode (L) within the photomultiplier tube amplifies the photocurrent to a measurable quantity. Normally, each photon produces 2.5 electron when it interacts with one dynode.
5. Analysis of negative voltage pulse gives the full energy spectrum of the radiation. A schematic representation of pulses at the probe unit (F), an amplifier (H), and a scaler (J) are shown in Fig. 6.1.

6.6   Scintillator

Type of Scintillator (Phosphor Materials) Needed for Measuring the Activities of Radioactive Isotope

γ-rays emitted by an isotope, when interacting with a phosphor, they excite atoms to a higher energy. When these are de-excited, photons are emitted. Therefore, phosphor material should easily get excited and quickly de-excited. Moreover, excitation process should occur with radiation of the lowest possible energy, so that even low energetic radiation can be detected. In addition, wavelength of the photons emitted during de-excitation process should be in the visible region, i.e., in the range of 400–600 nm so that they could be registered by the photomultiplier. Considering these requirements, we are left with only a few types of phosphor materials. The characteristic properties of some of the suitable materials are given in Table 6.1.

These phosphor materials are also called **scintillator**. The photons emitted by phosphor during de-excitation are used by the photomultiplier to measure and identify the radioactive sample. Hence, scintillator is the main constituents of the scintillation counters. As suggested in Table 6.1, there are two types of scintillator; **inorganic** and **organic scintillator**.

Table 6.1 Some useful properties of few fluorescent materials normally used in scintillation counter

Materials	Density gcm^{-3}	Wavelength of maximum emission Å	Decay time of excited atom in sec
Liquid phosphors	0.86	3500–4500	2.8×10^{-9}
Plastic phosphors	1.06	3500–4500	3.5×10^{-9}
Anthracene crystal	1.25	4400	3.0×10^{-8}
NaI(Tl)	3.67	4200	3.0×10^{-7}
CsI(Tl)	4.51	4200–5700	1.1×10^{-6}

6.6.1   Inorganic Scintillator

Inorganic scintillator are high density materials. Probability of interaction of γ-rays with phosphor increases in the presence of high density material. Therefore, inorganic scintillators are generally iodide salts of alkali metals as iodine atoms are the largest anions offering a large area for stopping γ-rays. Thallium ion is used as a dopant in these crystals because it has the property to shift the wavelength of the photon into the visible region. The most common scintillator is NaI(Tl), but CsI(Tl) is also used in some special cases (Table 6.1).

Inorganic scintillators are designed to give either a flat (Fig. 6.2K) or a well type surface (Fig. 6.2L). Flat surface scintillators are generally used for counting solid sample. They give about 2π-geometry. For counting liquid sample, a good type of crystal is used. A test tube made of transparent glass, containing the radioactive sample, is inserted into the well of the scintillator (L) for counting. This method can give nearly 4π-geometry for counting because all radiations, except those traveling in upward direction can penetrate through glass of the sample holder tube to reach the NaI(Tl) phosphor. However, radiations traveling in upward direction are lost and not counted.

There are different sizes of inorganic scintillator, having 50 mm, 75 mm, and 100 mm diameter. Diameter of the photomultiplier tube should be same as that of the crystal. Larger the size of the scintillator greater is the chance for γ-rays interaction and larger is the efficiency of counting. However, background count increases with size of the crystal. Price of large scintillator is also very high. The most common sizes of single crystal used for this purpose have 2" and 3" diameter.

Alkali halide scintillators are hygroscopic, and hence, they have to be protected from air and moisture. All surfaces of single crystal except the bottom side are coated with a thin layer of Ag_2O or Al_2O_3. Then, the entire surface except the bottom surface

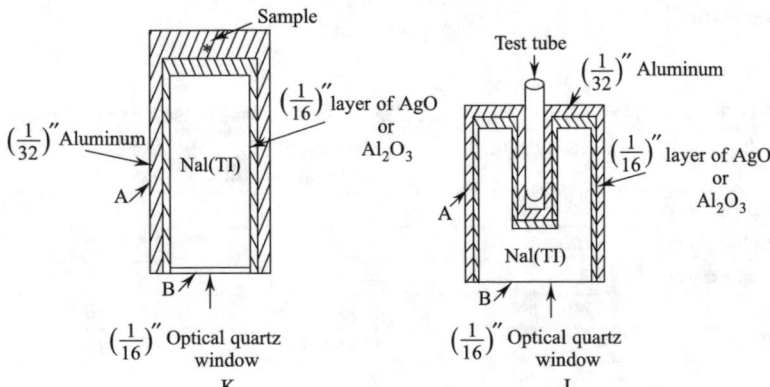

Fig. 6.2 Schematic representation of NaI(Tl) single crystal scintillator: (K) NaI (Tl) with flat bottom and top for solid sample; and (L) NaI(Tl) with well to place liquid sample

is enclosed in a specially designed aluminum cover (Fig. 6.2). So far the bottom surface is concerned, i.e., the side (B) has to be exposed to the photomultiplier tube. Hence, this side should be transparent. For this reason, a quartz (optically flat) plate is used to seal the bottom of the crystal (B). Because of the aluminum cover (A) over the crystal, weak β-particles or α-particles cannot be counted. However, β-particles of only high energy can be counted by this scintillator, provided its energy is greater than energy required to penetrate the thickness of the aluminum casing.

6.6.2 Organic Scintillator

Organic scintillator is made of an organic compound, where the basic structure is made of carbon atoms. Thus, the probability of interaction of γ-rays with the scintillator is less and, hence, the efficiency of counting γ-rays with organic scintillator is low. However, β-particles and α-particles can be counted with a greater efficiency due to higher probability of their interaction with the scintillator.

6.6.2.1 Organic Liquid Scintillator

Organic liquid scintillator consists of an organic scintillator, a wavelength shifter, and a solvent. Particulate radiations (e.g., α-particles or β-particles), present in the radioactive sample, interact with the scintillator to produce photons. The wavelength shifter helps to shift the wavelength of the photons to a visible region. These photons then interact with photomultiplier tube, (as discussed earlier), to produce electrons which are counted and recorded by the method discussed earlier (Fig. 6.1). Organic scintillators are normally kept in colored bottle and are always kept under an inert atmosphere like nitrogen. The liquid scintillator should be kept in a dark bottle, nitrogen atmosphere, and at low temperature, otherwise scintillation properties gradually diminish. Since such scintillators are made of carbonaceous materials, they can be useful for counting low energy β-particles or α-particles.

6.6.3 Composition of Scintillator

In the liquid scintillation method, the radioactive material is homogeneously mixed with one of the scintillator listed below and is kept in a perfectly clean flat bottom glass vial. This vial is then kept over the photomultiplier tube using some grease to ensure a good contact of the bottom of the glass vial with the photomultiplier tube. Since the radioactive sample is homogeneously mixed with the of α-particles or α-particles or β-particles by the liquid is minimum. Thus, by this technique, liquid sample can be counted without making a dry source. The solutes and solvents which are used in a scintillator are listed below

1. Types of organic scintillator

    (a) *p*-terphenyl

    (b) 1,2-phenyl-(4-biphenyl)-oxazole (PBO)

    (c) 2-phenyl-5(4-biphenyl)-1,3,4-oxidiazole (PBD)

    (d) 1-4-di(2-(5-phenyloxazolyte))-benzene (POPOP)

2. Types of solvents

    (a) Xylene (b) Toluene

    (c) Phenylcyclohexane (d) Shellsol-A

    (e) Decaline (f) Dioxane

Xylene and toluene are the most common solvents used for counting samples dissolved in non-polar solvents. Dioxane is used for samples containing up to 30% water. Other solvents are used when large volume of scintillator is to be used.

6.7   Precautions Liquid Scintillation

The radioactive material to be counted, should be soluble in the scintillator, for which one has to use a suitable solvent. But one cannot use any solvent as there are some solvents capable to absorb the photons (released by the scintillator), e.g., acetone, halogenated solvents, alcohol, etc. One of the essential conditions for accurate measurement of radioactivity by organic scintillator is the presence of a homogeneous and stable mixture of radioactive solution and the scintillator. Adsorption of radioactive materials to the wall of the container or precipitation of radioactive material from liquid scintillator sample are the major problem with organic scintillator. These factors lead to an appreciable error in the determination of the activity of the sample. Moreover, the radioactive solution should be colorless, and should not absorb the photons released by the scintillator. However, it is possible to make a correction for the loss of count for absorption up to 1–10%.

6.7.1   Size of the Sample Holder

The volume of solution to be counted is also determined by the diameter of the photomultiplier tube. Minimum amount of liquid scintillator should be used. Usually, 1 ml scintillator, 1–2 ml of the radioactive sample, and 5–10 ml solvent are used for liquid scintillation counting. Volume of liquid for counting depends upon geometrical arrangement of the photomultiplier tube. When the photomultiplier tube is kept vertical (Fig. 6.3A, B) then the diameter of the bottom of the sample holder should be 1/10th smaller than that of the photomultiplier tube (Fig. 6.3A).

Height of the liquid in the sample holder should be about 1 cm. Diameter of the sample holder is kept smaller to minimize the amount of radiation escaping

Fig. 6.3 Schematic representation of photomultiplier tube (PM) and the sample: **A** vertical arrangement with diameter of the sample tube slightly smaller than that of the PM tube, **B** vertical arrangement with diameter of sample tube larger than PM tube, **C** horizontal arrangement with height of sample tube almost same as that of diameter of PM tube, and **D** horizontal arrangement with height of the tube larger than that of the PM tube

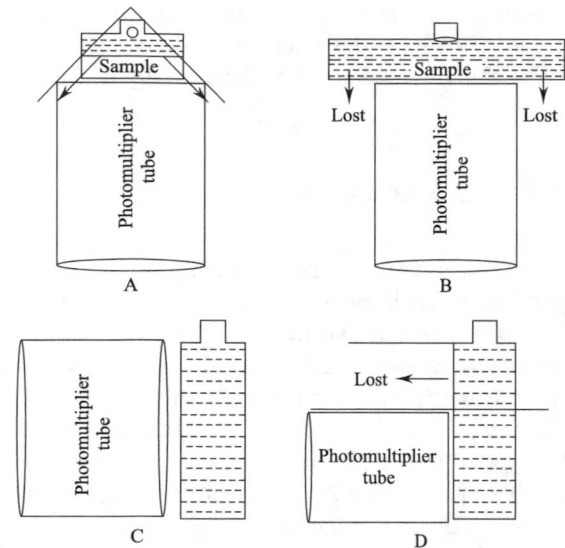

from the photocathode. If the sample holder has a large diameter (Fig. 6.3B), then radiations escape and do not reach the photocathode. The efficiency of counting, hence, becomes low in this case. Similarly, if photomultiplier tube is horizontal (Fig. 6.3C), then preferably, diameter of the sample holder should be 1/10th smaller than the photomultiplier tube, and height of the liquid should also be 1/10th less than height of the photomultiplier tube (Fig. 6.3C). It should not be of the size as shown in Fig. 6.3D.

6.7.2   Composition of Sample for Counting

The best resolution of counting is obtained when low concentration (5g of *p*-terphenyl/liter) of the solute is dissolved in toluene. A combination of polystyrene (97%) with a solute, *p*-terphenyl (9.5%) and a wavelength shifter, tetraphenyl butadiene (0.03%) has proved to be a very useful organic scintillator. One does not need to prepare these scintillators, as manufacturers supply specific scintillator dissolved in the required solvent. It is always better to ask the manufacturer for detailed information about the constituents of the scintillator before purchasing.

To achieve a geometry as shown in Figs. 6.3A and C, one may have to take 10–12 ml liquid. Composition of the liquid in the sample holder should be adjusted so that out of total volume (10–12 ml) it contains 1–2 ml of organic scintillator, 1–2 ml of liquid to be counted (there is normally no restriction to this amount), and remaining volume is made by a suitable solvent. However, users are expected to find out the best composition by a trial and error method, keeping in mind that a minimum volume of the liquid scintillator should preferably be used.

6.8 Gel Scintillator

Many times measurement of activity of ^{14}C or ^{3}H or any other low energetic β-particle present in an inorganic substance, which are insoluble in non-polar solvents like toluene or xylene is required. Moreover, the substance may be insoluble in non-quenching solvents. Such materials, however, can be counted by a gel-type scintillator. This scintillator consists of a mixture of solid organic scintillator powder and silica gel. The solid sample is thoroughly mixed with the gel scintillator. Sometimes, it is better to put the container under ultrasonic vibrator for thorough mixing of sample with the scintillator. The glass vial containing this mixture is heated to about $50°$–$60°$ C for few a minutes. During this process, the mixture forms a transparent gel with solid radioactive sample dispersed homogeneously. On slow cooling (i.e., up to room temperature), the container of the glass vial becomes almost a transparent solid gel. By this process, we get the radioactive sample thoroughly dispersed in the solid suspension of transparent silica gel. The glass vial is then put on the photomultiplier tube, as described for NaI(Tl) system and the radioactivity is counted as discussed earlier. Size and shape of the glass vial depend upon size of the photomultiplier tube.

Manufacturers of scintillator, however, keep on developing new kinds of scintillator, and it is thus always advisable to ask for details of all types of scintillator manufactured by the various firms so that one can select the best-suited scintillator for the experiment.

6.9 Filter Paper Soaked with Scintillator

One can also use low carbon content filter paper for counting purpose. The sample (in slurry form) is homogeneously spread over a filter paper and connected to a small vacuum system. The solution is then filtered through the filter paper and semidried by sucking air for some time. This paper along with the thin film of solid radioactive sample is carefully transferred to a counting vial, such that the filter paper sits on the bottom of the vial with radioactive sample facing upwards. Liquid scintillator and solvent are added from the side of the wall of the vial, taking care that the thin layer of radioactive sample is not disturbed. Total volume of the liquid is maintained to about 10–12 ml. Filter paper in the presence of liquid scintillator becomes transparent. Radiations emitted by the sample interact with the liquid scintillator producing photons which interact with the CsI photocathode of the photomultiplier tube. For this purpose, the vial is kept over the photomultiplier tube, with the help of silicone grease. The assembly is then set, as shown in Fig. 6.1, and counting is done, as described earlier.

This method is used for samples, which are colorless and insoluble in solvent (which does not quench). Preferably, amount of solid is kept to a minimum quantity in order to minimize the loss of activity due to self-absorption of radiation by the sample.

6.10 Parameters Controlling The Scintillation Counter

In counting a radioactive sample containing a γ-emitter or an α-emitter usually one has a dual interest. *Firstly* identification of radioactive material, by knowing its energy spectrum of the radiation. *Secondly*, measuring total activity either at the photopeak ($A1$, Fig. 6.4) or whole photoelectric spectrum (area under curve A) or entire activity inclusive of Compton scattering effect (in case of γ-rays), i.e., entire area of spectrum from C to D (Fig. 6.4). For these purposes, characteristic of scintillation counter needs to be established, which depends upon nature of the radioactive sample to be counted, i.e., whether it is a β-emitter or an α-emitter or γ-emitter. Conditions for counting these radiations are hereby briefly explained.

6.10.1 γ-radiations

Conditions for counting a sample containing γ-emitter depend on whether the counting required to measure the activity corresponding to its intensity of entire area of the photopeak (i.e., area under the photopeak A or B of a typical γ-spectrum of a sample as shown in Fig. 6.4) or at the photopeak (i.e., at the peak of spectrum A or B) or

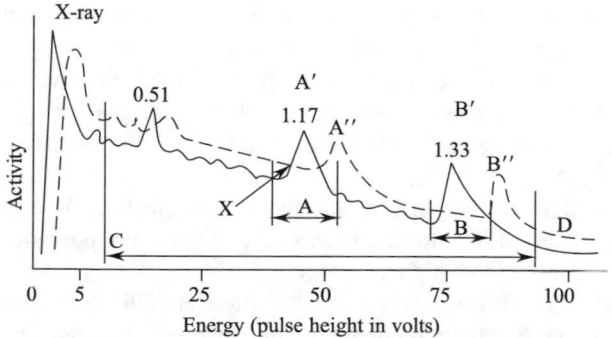

Fig. 6.4 A typical schematic spectrum of a α-emitter (full line) showing the effect of voltage fluctuation on the shift of spectrum (broken line), A and B represent the total area of the two photopeaks A' and B', respectively, and A'' and B'' show the shift in the position of photopeaks

the total γ-rays spectrum to be counted (i.e., area under the spectrum covering area from C to D of Fig. 6.4). For either of these conditions EHT, a suitable amplification and suitable channels of a pulse height analyzer are to be selected.

6.10.2   β-particulate Radiations

Optimum EHT and optimum amplifications are to be found out to cover the full β-spectrum (i.e., either C or D of Fig. 5.3) such that the entire spectrum lies within 0–100 V range of the pulse height analyzer.

6.10.3   α-particulate Radiations

Optimum EHT and the optimum amplification are to be found before counting α-particulate radiation at either the photopeak (i.e., at X in Fig. 5.3B) or entire area of the photopeak (i.e., the area of spectrum between K and L Fig. 5.3A). It may be noted that with α-counting, normally one does not get an X-ray or Compton scattering radiation (as obtained with γ-rays). As a result, background counts can be eliminated very easily with pulse height analyzer, because with pure α-emitters, pulses of α-particle would be much bigger than that pulses produced by the background radiations (Fig. 5.11).

6.11   Optimum Conditions for Counting

In either of the cases mentioned above, EHT applied to the photomultiplier tube and amplification factor are set, such that the entire spectrum is within 0–100 V of the pulse height analyzer limit. No hard and fast rule stands to achieve this setting, it comes only by practice. Usually, rate meter (which gives an average count rate) is used instead of scaler. A particular EHT is applied (usually a lower value) and the entire range of the pulse height is scanned by altering the pulse height gate from 0 to 100 V. Count rate in the rate meter is recorded to check the nature of the spectrum of radiation (Fig. 6.5).

For example, if the entire spectrum is found to lie within 0–20 V range of the pulse height analyzer, then EHT is further increased and the entire operation is repeated. This shifts the spectrum to a higher voltage.

If the rate meter shows low count rate, then amplification factor is further increased. However, if activity recorded by the rate meter does not show much variation in counts for the entire range of pulse height (i.e., 0–100 V) and the count rate is very high, then the amplification factor is reduced. By this trial and error method eventually one gets a suitable EHT and amplification value to get a spectrum

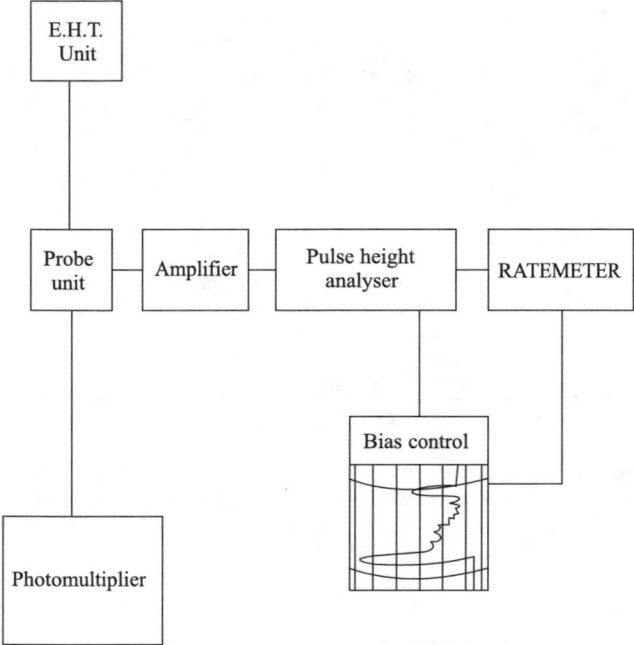

Fig. 6.5 A typical setup for α-ray counting

as shown in Fig. 2.3 (for γ-rays) and Fig. 2.4 (for β-particulate radiation) or like the hypothetical Fig. 6.5 (which shows spread spectrum throughout the pulse height gate. Generally, an increasing voltage to photomultiplier tube spreads the spectrum (i.e., lengthwise) and an increase of amplification increases the height of pulses. γ-spectra of Manganese-54 is shown in Fig. 6.6 where one can see the impact of EHT and amplification on the spectrum. The best spectrum for this isotope is observed at 1015 V with 8×500 gain of amplification. Any other setting does not give satisfactory photopeak position within 0–100 V range of the pulse height analyzer.

6.11.1 Calibration of Pulse Height Analyzer

If the range of pulse height analyzer (0–100 V) is calibrated in terms of energy (MeV), position of the photopeak (volts) can be assigned a particular energy value and thus the isotope can be identified. The calibration of pulse height analyzer is done by recording γ-spectrum of four to five known isotopes giving different γ-rays with different energies. The voltage of the pulse height analyzer corresponding to the energy of γ-rays recorded by the scintillation counter is calculated from the recorded spectra for each sample. Finally, energy of γ-rays is plotted against the voltage of the pulse height analyzer corresponding to their photopeak. This gives a linear graph

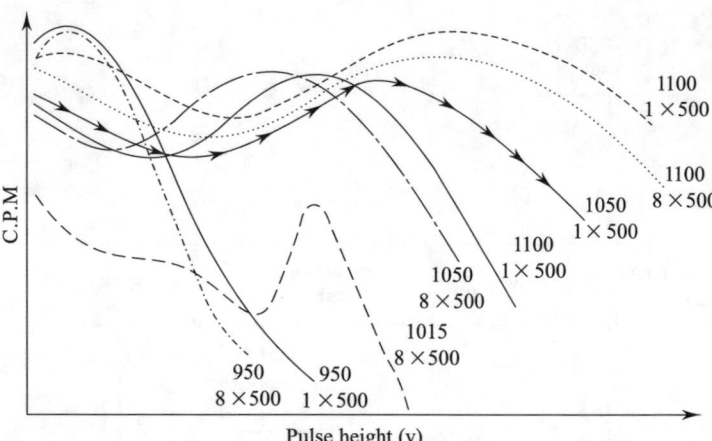

Fig. 6.6 A typical graph obtained with Manganese-54 isotope with NaI(Tl) scintillation counter showing the effect of variation of amplification factor and EHT on count rate versus the pulse height analyzer voltage. Amplification factors, e.g., 1 × 500 and 8 × 500 are shown below the EHT voltages, e.g., 950, 1015, 1050, 1100, and 1015 V

which can be used as a calibration graph of pulse height analyzer. It should be noted that for this experiment one should not change EHT and amplification factors while taking γ-spectra of different γ-rays. However, this calibration is valid only for conditions under which the spectrum is recorded. If any value of amplification or EHT are changed, calibration of the pulse height analyzer has to be observed again (obvious from Fig. 6.6). When the values of EHT and amplification for counting a particular isotope are confirmed by the above procedure, the radioactive element can be identified by knowing the energy of γ-rays. Activity of the sample can be measured by counting the sample by any of the following procedures:

- Pulse height analyzer can be set to a voltage corresponding to the peak of the pho-toelectric spectrum with a window of ±1.0 volt (i.e., either at A' or at B' Fig. 6.4). Under this condition, counting rate is low with almost negligible background count rate. However, counting at the photopeak has a disadvantage, specially if the supply of voltage to the unit fluctuates. Such fluctuation drifts the photopeak from A' to A'' or B' to B''. If such a shift occurs, total count recorded by the scaler would decrease, as the system would be counting either at A'' or B'', depending upon setting of the pulse height analyzer. Therefore, it is always better to operate the instrument with a stabilized power supply. In the absence of a stabilization unit, if counting is performed for a series of samples, where it is expected to observe a continuous or abrupt changes in the activity, it would be difficult to confirm whether change recorded activity is due to change in the activity of sample or shift in the photopeak due to the fluctuation in voltage supplied to the electronic units or it is due to both.
- One can count the entire photopeak, either A or B by setting not only the pulse height analyzer at voltage corresponding to the photopeak A' or B' but also the

window width such that it covers the area of the photopeak (A or B), because
window width can set the lower and upper value from the photopeak value [i.e.,
$A' \pm$ (widow width)]. This method reduces background activity to about 20–40
cpm and can tolerate, to some extent, small fluctuations in the voltage change.
Nevertheless, it is advisable to use voltage stabilized units for this method also.

• Pulse height analyzer could be used as a discriminator bias where voltage in the discriminator is set at a value corresponding to the starting of the photopeak (Fig. 6.4).
Under this condition, all pulses greater than the set voltage would be recorded.
For example, if the discriminator bias voltage is fixed at potential corresponding
to the beginning of the photopeak, the counter will record all pulse greater than
this voltage to its upper limit (i.e., 100 V). The advantage of this method is that
this type of counting is not affected by any fluctuations in power supply as it did
while counting at the photopeak. Moreover, if the activity present in the sample
is very low to be detected effectively at the photopeak, then also this method is
preferred. But the background count under this condition would be much higher
than what one would get by counting the sample at the photopeak. Normally, the
background count under this condition is around 100–200 cpm.

6.12 β-counting

Since β-particles, unlike γ-rays (Fig. 2.3), have a continuous spectrum (Fig. 5.3C),
counting condition for such particles are different from that of γ-rays. In fact, there
is no hard and fast rule to find out the best operating condition, but the following
method can be adopted to find out the best counting condition for a sample containing
β-particles.

6.12.1 Counting Conditions for β-particles

An amplification factor in the amplifier is set to an arbitrarily convenient value to
start with. Then, the activity is measured for various applied EHT, and a graph is
plotted between the count rate and the EHT voltage (Fig. 6.7).

This procedure is followed to count radioactive sample (S) as well as background
activity (B) (i.e., counting without any radioactive material in the counter), separately (Fig. 6.7). The pulse height analyzer is used as a discriminator and its gate
is selected at a voltage, at which all X-rays peaks are cut off (i.e., voltage corresponding to position marked "C" in Fig. 6.4). This is normally achieved by setting
the discriminator value at 5–10 V. In Fig. 6.7, S is the usual type of curve obtained
with any sample containing β-emitter, and B is the activity due to the background
activity. Figure 6.7 shows that a certain potential activity due to the background is
greater than that of the source. One should not be alarmed by this observation. The
reasons for getting such a result is not yet fully understood. In Fig. 6.7, similar to an

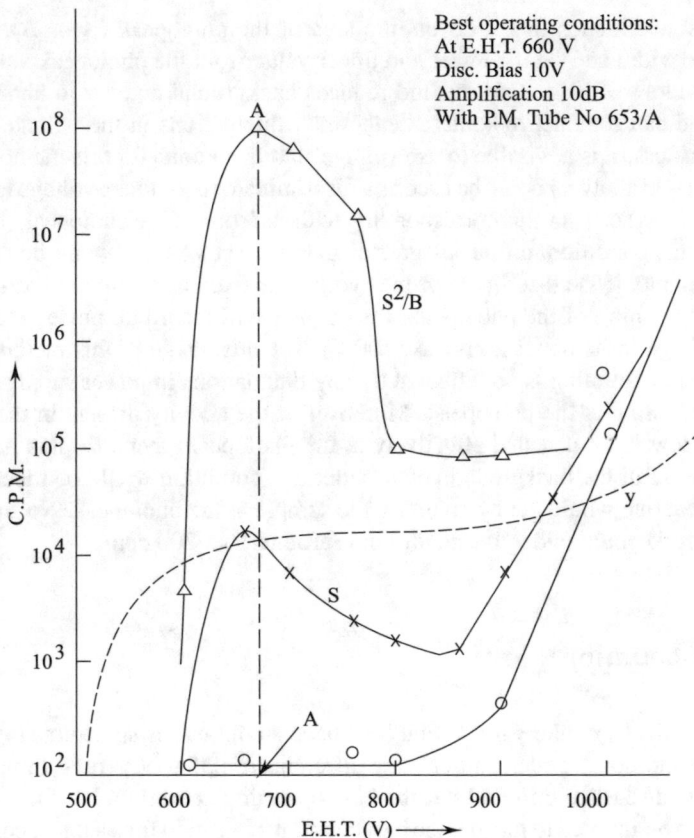

Fig. 6.7 A typical graph showing variation of count rate of Chlorine-35 isotope (S) and the background count rate (B) versus applied EHT to the photomultiplier at a discriminator bias of 10 V. The plot of S^2/B versus EHT calculated from the plot of S versus B shows a maximum (A) at EHT 660 V. Sometimes a plateau instead of maxima is also observed with scintillation counter for β-particles (Y)

ionization counter, plateau value does not appear (i.e., for the source S), and thus it becomes difficult to suggest the operating potential for counting the isotope for which this type of graph is obtained.

When no plateau is observed (as shown in Fig. 6.7), one usually plots S^2/B against the applied potential (EHT voltage), which results in a parabola. EHT potential corresponding to the maxima (A in Fig. 6.7), is taken as the best operating potential because at the maxima of parabola there would be maximum difference between the activity due to the sample (S) and the background (B). If the parabola is not obtained then amplification factor is altered and the entire counting operation is repeated, till one gets a good parabola. However, with β-particles, sometimes, one can get a plateau, shown by broken line in Fig. 6.7Y.

It is worth remembering that the characteristic obtained for β-counting by the scintillation counter is not as universal as one gets with G.M. counter. The settings obtained with the scintillation counter correspond to the isotope for which the particular amplification and EHT settings have been observed. When isotope or amplification or EHT values are changed, maxima of the parabola also changes. Therefore, it is necessary to find the best operating condition, i.e., the parameters like amplification and EHT for the β-particles emitted by the radioactive isotope to be counted.

6.13   Quenching Corrections

In liquid scintillation counting, one has to ensure whether sample or solvent absorbs visible photons produced by interaction of radiation with scintillator. As mentioned earlier, all colored samples, water, halogenated solvents (CCl_4, $CHCl_3$, etc.), alcohols, acetone, etc., are known to absorb visible photons produced from the scintillator. These chemicals/solvents are designated as **quenchers**. In order to appreciate the effect of quenching on the spectrum of β-particles, spectrums of β-particles of Chlorine-36, (LiCl labeled with Chlorine-36) dissolved in toluene (Fig. 6.8A) as a solvent and in acetone as a solvent are shown in Fig. 6.8B. These spectrums are taken under similar condition (i.e., at the same operating conditions of the scintillation counter). It can be seen that the addition of acetone and water reduces β-spectrum as well as shifts the whole spectrum toward lower energy. These two figures suggest that it is essential to find out the exact percentage loss of activity due to the quencher (acetone/water mixture) for getting an accurate activity of the sample. Many methods have been devised to correct the count rate for loss of the activity due to such quenching effects. Among these, some of them are discussed here.

One of the method of quenching correction is by drawing a graph of percentage loss of sample's activity against the volume of quencher added to the solution containing known but fixed quantity of the radioactive sample. A typical nature of such graph showing the decrease in activity recorded due to the presence of quencher is shown in Fig. 6.9.

This graph can be used for quenching correction. By knowing the volume of quencher added into counting sample, the corresponding percentage loss of activity due to quencher can be calculated from the graph shown as f_1, for example, of 0.2 ml of added quencher (Fig. 6.9). Water was used as a quencher solvent for this experiment. This correction is valid only for the isotope for which this type of graph is drawn and for operating condition of scintillation counting system. If there is any change in the isotope, solvent or operating condition, a fresh quenching correction has to be determined. This method is also known as **internal standard technique**. Another method for quenching correction is when the amount of quencher present in the counting sample is not known (known as **external standard method**). Both these methods are discussed in detail in the foregoing sections.

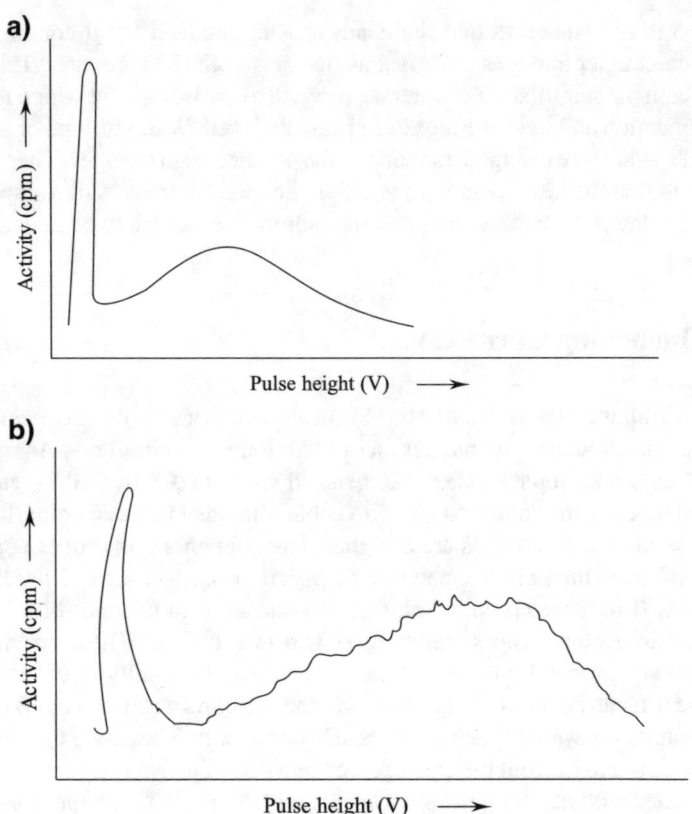

Fig. 6.8 A Effect of quenching on the activities. β-spectrum of ^{36}Cl, using liquid scintillation counter (EHT 660 V, disc. bias 10 V and amplification 10 dB). The solution contains, toluene (2.0 ml), aqueous solution of LiCl labeled with ^{36}Cl (0.0001 ml), liquid scintillator (5.0 ml). **B** Effect of quenching on the activities. β-spectrum of ^{36}Cl, using liquid scintillation counter (EHT 660 V, disc. bias 10 V and amplification 10 dB). The solution contains acetone (1.0 ml), aqueous solution of LiCl labeled with ^{36}Cl (0.0001 ml), liquid scintillator (5.0 ml) and 1.0 ml water

6.13.1   Internal Standard Technique

For this method, a radioactive source (e.g., ^{36}Cl, ^{14}C) of known activity (i.e., of known disintegration per unit time, e.g., X dpm) is dissolved in a solvent which does not quench (or at least the amount of quencher is negligible, to have any effect on counting). The activity of this sample is measured (Y cpm) with a liquid scintillation counter. From these two measurements, the efficiency of counting is calculated which is equal to ($\frac{Y}{X} \times 100$). Then a small known amount of quencher (methanol, water, etc.) is added to the sample, and percentage loss of activity is calculated from the observed count rate. This process is repeated by adding different amounts of the same quencher to the same sample, and % efficiency of counting is calculated for

Fig. 6.9 A typical graph showing gradual decrease in activity of the sample due to addition of a quencher. The LiCl labeled with ^{36}Cl (0.10 mC dissolved in 0.01 ml of HCl) was used for this experiment

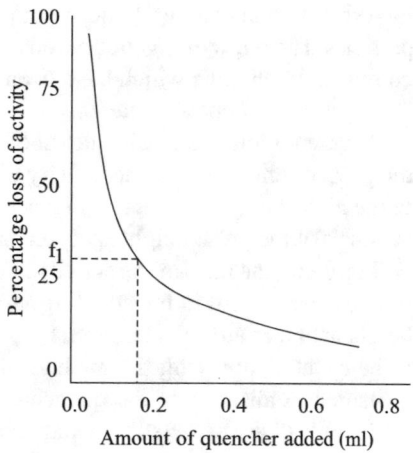

each case. Finally, a graph is plotted between percentage efficiency of counting observed with each added quencher versus the amount of quencher added (Fig. 6.9). However, if disintegration per unit time is not known, then the activity measured in absence of any quencher can be taken as initial count rate and percentage loss of activity is calculated for each added amount of quenching solvent. Finally, a graph (Fig. 6.9) is plotted between percentage loss of activity of the sample versus the amount of quencher added into the sample. This graph can be used for calculating loss of activity due to the quencher present in the sample. Detailed procedure of this technique is briefly discussed here.

Activity (A) of the sample (for which quenching correction is to be made) containing known amount of quencher is recorded under identical counting condition (i.e., condition under which Fig. 6.9 was drawn). With the help of Fig. 6.9, percentage loss of activity for the corresponding amount of quencher present in sample-A is found (call this factor "f_1"). This factor is multiplied with the activity (A) to get true activity of the sample (i.e., actual activity, had there been no quenching effect).

This method is, however, applicable only when the amount of quencher present in the sample is known. Moreover, those quenchers which reduce the activity to 50–60% by addition of 2–3 drops of quencher (e.g., methyl iodide labeled with ^{14}C as solute), is not suitable because a large error is involved in getting the correction factor "f_1". For such cases, an external standard method is used.

6.13.2   Channel Ratio Technique

The principle of this method is based on the following factors. We have seen earlier that the spectrum of β-particle contains about 60% of total β-particles, with energy corresponding to 1/3 of E_{max} (Fig. 2.1). Therefore, whenever β-particles are counted by a scintillation counter, majority of photons are generated due to lower energy β-particles. These low energetic photons are absorbed by the quencher more effectively, compared to photons which have been produced by higher energy β-particles (i.e., by β-particles of energy near E_{max}).

Thus, addition of a small amount of quencher in the counting sample compresses the β-spectrum such that the activity of lower β-particle is reduced more compared to the high energy β-particles (Fig. 6.10). If activity of β-particle is recorded in two channels of the pulse height analyzer, say in channel corresponding to energy A and B (Fig. 6.10), the ratio of the activity in the two channels will change with the amount of quencher present in the counting sample. The channels "A" and "B" of the pulse height analyzer are selected first, by plotting β-spectrum of the radioactive isotope to be counted, applying the method discussed earlier. In other words, spectrum is measured in absence of the quencher. Two channels (i.e., pulse height voltages) "A" and "B" are selected by examination of this spectrum, such that, the activity at one pulse height setting bias (A) is about 80% of that recorded at the pulse height setting (B). Now for performing the experiment, pulse height analyzer is used as a discriminator. Two values of the discriminating bias (or gate) are selected to correspond to the potential at "A" and "B" position of the spectrum. Sample is counted at these two values of discriminators separately. In some sophisticated liquid scintillation counting systems, arrangements are made internally to give count rates at two values of the pulse height voltages, so that activity in one channel is 80% of the other channels.

Quenching correction can then be carried out in the following manner. A β-emitter sample, of known activity, in absence of any quencher is counted to get the sample's detection efficiency by using the method discussed in the previous section (Fig. 6.9). Let a radioactive sample (S) giving disintegration per minute (Y dpm) be recorded in channel A giving an activity of x cpm. Then, its detection efficiency "f" in channel A is given by

$$f = \frac{x/\text{cpm} \times 100}{y/\text{dpm}} \tag{6.1}$$

The activity of this sample (S) is also determined in channel B. After this counting, a small amount of quencher is added to the sample (S), and its activity in channel A and B are determined. In this fashion, the experiment is repeated for various amount of the quencher added to the sample (S). Detection efficiency in channel A for each sample is calculated from Eq. (6.1). Finally, a graph is plotted between the ratio of activity recorded in channel A and B versus the detection efficiency in channel A calculated from Eq. (6.1). This graph is expected to be almost linear for small amount of added quencher (i.e., when the activity does not fall down due to the addition of quencher) to less than 20–30% of the initial activity (Fig. 6.11).

Activity of the sample (whose amount of quencher present in the mixture is not known) is now recorded in channels A and B. The ratio of activities in channels A and B is calculated (assume it to be K_1). With the help of the previously calibrated graph (Fig. 6.11), detection efficiency of the sample for the corresponding value of

Fig. 6.10 A typical β-spectrum of a radioactive sample (X) in absence of any quencher (shown by full line). The spectrums shown by broken lines are for the same sample but with the presence of increasing concentration of quencher (a, b and c). A and B are two selected channels such that activity in channel A is 80% less than that of channel B

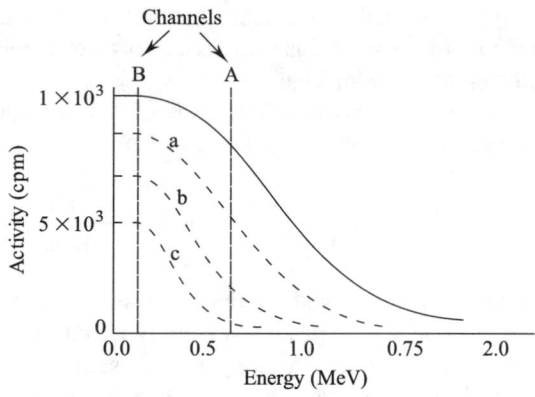

Fig. 6.11 A typical graph showing variation in the ratio of activities recorded in the two channels (i.e., ratio of activity in channel "A" and activity in channel "B") versus the detection efficiency recorded in channel A for various amount of quencher added to the same radioactive sample

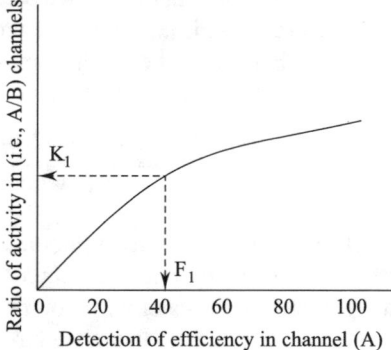

K_1 is observed which is F_1, for example. The activity calculated in channel A is then multiplied with this detection efficiency factor F_1 to get corrected activity of sample containing the unknown amount of quencher.

This method, however, is not useful, when quencher reduces the activity to less than 5–10% for addition of even 0.5 ml to 15–20 ml solution of radioactive sample. Such solvents should be avoided in scintillation counting. It is worth mentioning again that this method is applicable provided we know the type of quenching liquid present in the counting samples. Without this information, the standard graphs (Figs. 6.9, 6.10, or 6.11) cannot be drawn. This problem arises as quenching can occur due to any of the constituents of the liquid sample to be counted by the liquid scintillation counter, e.g., the solvent, the solute, or both.

6.13.3 External Standard Source Techniques

Sometimes, the external standard source technique is also used for calculating the counting efficiency of sample containing either colored sample or unknown amount of quencher. In this technique, the effect of superimposition of external radiation on activity of the sample is measured in both channels. From these data, the efficiency of counting is calculated.

A sample of know activity is recorded in channel A (let it be x_1 cpm), and its detection efficiency is calculated as before. i.e.,

$$f = \frac{x_1 \times 100}{x_{dpm}} \qquad (6.2)$$

where x_{dpm} is expected activity of the sample in terms of disintegration per minute. An external source of known activity is put on the top of sample tube (not in the solution) and its activity along with the sample is recorded in channel A (let it be x_2 cpm). The increase in activity due to external γ-source ($x_2 - x_1$ cpm) is calculated. Then required amount of the quencher is added to the sample and activity in channel A is recorded, with (x_i cpm) and without (y_i cpm), in the presence of γ-source. Finally, relative detection efficiency in channel A is calculated by equation given below

$$v = f \times \frac{\text{Activity with quencher in channel } A}{\text{Activity without quencher in channel } B} = f \frac{x_i}{y_j} \qquad (6.3)$$

where v is the relative detection efficiency. A graph is plotted between the relative detection efficiency v and increment in activity of the sample (i.e., $y_i - x_i$) due to the presence of the external γ-source.

The sample with an unknown amount of quencher is counted with (z_1 cpm) and without (z_2 cpm) the γ-source. Increment in activity ($z_2 - z_1$) due to the external γ-source is calculated. By using the standard graph (Fig. 6.12), relative detection efficiency (f_1) is found for the corresponding increment ($z_2 - z_1$) in the count rate observed for the sample. The corrected activity for quenching is calculated by multiplying relative detection efficiency (f_1) with count rate, recorded in channel $A(z_1)$ for the sample without external γ-source, i.e., product of z_1 cpm and f_1. In this method, it is necessary that external γ-source be kept exactly at the same position every time, and there should be no change in activity of γ-source with which these

Fig. 6.12 A schematic graph showing relationship between the relative detection efficiency and the increment in the count rate due to the presence of external γ-source

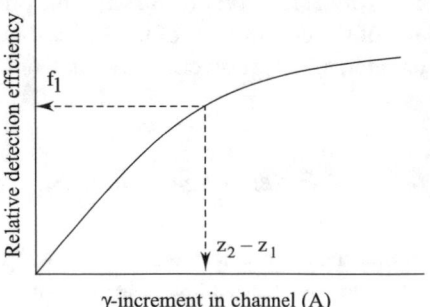

measurements are being made (i.e., one should use a long-lived energetic γ-emitter of a life greater than one week at least). This technique is very useful in making correction for quenching due to colored samples.

6.14 Effect of Multiple Type Radiations on Counting

Most of the radioactive isotopes decay by emitting more than one type of radiation (Fig. 2.7). For example, ^{60}Co, decays by emitting β- and γ-rays. Should we set the counter to measure either γ-rays or β-particles of this isotope? Is it possible to set the counter such that we can count the activity due to one of these radiations such that the effect of other does not influence the total activity? Because, for example, when γ-rays are counted then β-particle will also get recorded, unless some special effort is not made to prevent β-particles entering the counter. If no special effort is made to prevent β-particles entering into the counter, then the percentage contribution of β-particles into the recorded count rate would depend upon the nature of setting of the counting condition. In counting of such isotopes, if the instrument is set at the photopeak, then the contribution due to β-particles would be negligible. Whereas, if setting is made to count the entire spectra of γ-radiation, contribution due to β-particle would influence the counting rate considerably. In the latter case, there is likelihood of interference of β-particles unless, effort is made to isolate the penetration of β-particles into the counter. However, it is not feasible with liquid samples using liquid scintillation counting system.

For liquid sample, elimination of β-particle can be done by counting with solid phosphor (NaI) counter by setting to count at the photopeak of γ-ray. Now, we shall discuss various conditions of counting to measure the activity of isotopes which emit more than one type of radiation.

- The influence of β-particle can be reduced by placing a metallic screen (which would stop β-particles but not the γ-rays) between the source and the detector. This screen is made of a low atomic number substance to cut down the external Bremsstrahlung radiation, as well as β-particles. However, effect of Bremsstrahlung's radiation can be eliminated by the pulse height analyzer, because they appear as an electromagnetic radiation of low energy and low intensity.
- The influence of β-particles can be eliminated by allowing it to annihilate near the source, so that the latter may be considered responsible for getting 0.511 MeV photon emission in the β-spectrum. This is accomplished by placing the source in between two aluminum plates thick enough for the positrons to lose all their energies. Under these conditions, emission of 0.511 MeV photons would be observed. By making the appropriate setting in pulse height analyzer, the effect of 0.511 MeV photon can be eliminated.
- For materials which decay by γ-rays, the emitted photons compete with internal conversion process. For radio-nuclides of low or medium atomic number elements,

X-ray photons produced by the internal conversion can be forcefully absorbed by a suitable screen separating the source from the detector.

- If radio-nuclide decays by emitting several γ-rays, it is necessary to account for each of them and correct the count rate obtained for any coincidence of these two γ-rays (as is the case with Cobalt-60 isotope). All these methods need to be applied only if absolute count rate is required. When relative activity of the samples are required, like in study of the rate of adsorption of a radioactive isotopes on a substrate, finding activities of two layers in the solvent extractions, etc., then there is no need to carry out these corrections, as the contribution of other radiations would be same in all the samples.

Summary

In this chapter, we learnt about the principle of scintillation counter. We learnt that photomultiplier tube is used to measure the photons emitted by the scintillator when particulate or electromagnetic radiations interact with phosphor. Phosphor used for γ-rays counting is normally NaI(Tl) and for particulate radiation either liquid or solid organic phosphors are preferred. We also learnt about the types of liquid scintillator and their composition and effect of sample size on the counting efficiency. Some special type of scintillator like gel scintillator and counting by using filter paper have also been discussed. Precautions needed in counting liquid samples, corrections for chemical, which quenches photons produced during scintillation, are also emphasized. Finally, precautions needed in counting radioactive isotopes emitting more than two types of radiations are also discussed.

Chapter 7
Non-conventional Detection Techniques

7.1 Introduction

This chapter is devoted to discussing a few special types of counters and techniques for the measurement and detection of radioactive isotopes present in either small quantity (for example, radioactivity present in food and measurement of activity slightly greater than background activity or low activity of α-particulate radiation).

7.2 Semiconductor Detector

After the development of the theory of $p-n$ junction, a wide variety of detectors have come up, which use semiconductors like silicon and germanium single crystals. These counters have attracted scientists because of their high efficiency and high resolution to count lower energy particulate radiations as well as soft X-rays. It would not be possible to discuss in detail the theory of semiconductor detectors, as it will form a book by itself. Efforts will be made to explain the principle in an easier manner to acquire the working knowledge of this detector.

7.3 Principle of Semiconductor Detectors

To understand the principle of a semiconductor detector, knowledge about it is required.

What is a band gap in a semiconductor? When a material crystallizes, its atoms come very close to each other so much that the electrons of the neighboring atoms undergo an interaction in three dimensions. For example, when silicon crystallizes into a solid form, electrons of each silicon atom interact with the electrons of other silicon atoms. These interactions cause some uphill in the initial energy of electrons

© The Author(s), under exclusive license to Springer Nature Switzerland AG 2021   119
M. Sharon and M. Sharon, *Nuclear Chemistry*,
https://doi.org/10.1007/978-3-030-62018-9_7

of each silicon atoms. Because of this interaction, energy levels of each electron of different atoms take two new energy levels, valence and conduction energy levels.

Why do they form two new energy levels? Electrons of an atom in isolation experience one kind of electrical force, which is due to the interaction of a positively charged nucleus with electrons as well as electron–electron interaction of the same atom. The energy of an electron of this isolated atom settles down to one equilibrium energy level. At this stage, the electron presumes that in its universe, there exist only one positive charge and its negative charge. This situation changes when more than one atom comes close to each other. Under this condition, electrons of both the atoms are surprise to experience the presence of another positively charged nucleus surrounded by other sets of electrons. This situation forces electrons to re-adjust their energies into two different energies levels: one higher than its initial value and the other lower than its initial value. For example, the reader of this chapter might be engrossed in reading this chapter, and their mind might have attained an equilibrium energy level. But as soon as you come to know that there is a very beautiful girl standing next to you, then certainly you get perturbed. If we can alter our equilibrium levels in such type of situation, then why not electrons?

In other words, when one electron of one atom comes close to an electron of another atom, they create two new energy levels (Fig. 7.1a): one above the initial energy level and the other below its initial energy level. Suppose we increase the number of electrons by increasing the number of atoms and allow them to interact, then each electron of each atom will likewise create two new energy levels. If there are N number of electrons from N number of atoms interacting, then there would be $2N$ number of new energy levels generated (Fig. 7.1b): "N" set of energy levels below their initial value (known as valence levels) and "N" set of energy levels

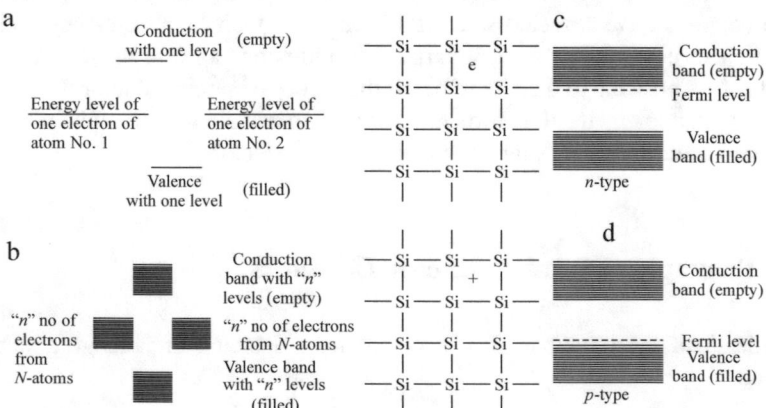

Fig. 7.1 A schematic diagram showing: **a** formation of two energy levels when two electrons come close together for interaction, **b** formation of N new energy levels when N number of electrons from N atoms come close enough for electron–electron interaction, **c** formation of N energy levels forming conduction and valence bands with n-type material

above its initial energy (known as conduction levels). If the magnitude of "N" is the Avogadro number, i.e., 6.023×10^{23}, then that many number of levels will be created in valence and conduction energy levels. Incidentally, we know that the chemical binding energy to hold a molecule stable is around 1–5 eV. Hence, as a crude approximation one can say that the electrons responsible to form the molecule must be occupying their energy within this energy value (i.e., 1–5 eV).

In other words, the depth of the "N" energy levels which we have generated should be related to this chemical binding energy. This means that 10^{23} energy levels lie within 1–5 eV depth. Hence, the difference between each energy level present within either conduction or valence energy would be $\approx 10^{-23}$. This value is so small that it becomes very difficult to differentiate one energy level from another (Fig. 7.1c). Hence instead of calling them a level, we call them a band of energy levels. The conduction and valence energy levels are recognized as conduction and valence bands, respectively.

Where do these electrons after creating these two energy bands exit? All electrons involved in this type of interaction would try to occupy either of these two bands. Filling of electrons will start with the lowest energy of the valence band. Once this band is filled, then it will try to occupy the conduction band. We know that one energy level can occupy a maximum of two electrons (like one couple can have a maximum of one wife making the number two), so if there were "N" number of electrons interacting with a similar number of electrons from other sets of atoms, then these $2N$ number of electrons can easily occupy the valence band (which has "N" number of energy levels).

Why do they occupy the valence band first? When water is spilled over a table, does it fly in the air or go to a lower height? Naturally, it follows the path which needs the least resistance. Likewise, these electrons will tend to occupy the lower energy band first and if some electrons are still left, then they will occupy the higher energy band. Since a valence band has "N" number of levels with a capacity to accommodate $2N$ number of electrons, all electrons of the interacting atoms can find their home easily within the valence band and thus the conduction band will remain empty. Thus, the net conclusion of these discussions is that when two sets of atoms come very close to each other, the electrons of these atoms interact forming valence and conduction bands.

7.3.1 Formation of Band Gap

How much far apart can these two bands be found? It is very easy to visualize. When a beautiful girl comes near your place, the level of excitation, i.e., creation of a new energy level will certainly depend how close the lady is to you. The farther away the girl is, the least amount of energy difference will be created. So is the case with the atom. Closer the atom gets, greater is the interaction and hence larger is the separation of the two new energy bands. Therefore, depending upon the nature of crystallization (i.e., lattice dimension of the crystal), band separation would be observed. The difference between the upper energy level of the valence band and lower energy level of the conduction band is known as the **band gap** and is express in terms of "*eV*". Hence closer the atoms, greater would be the band gap. This suggests that every material exhibits different band gap as they crystallize into different but specific lattice dimensions.

7.3.2 Fermi Energy in a Material

What is the magnitude of energy of electrons occupying the valence band? This is a difficult question to answer, because it involves a huge amount of mathematics to arrive at some meaningful conclusion. But it can be confirmed that if the material is highly pure (i.e., intrinsic semiconductor), the maximum energy electrons can occupy being in the valence band at room temperature is equal to half of the band gap. In other words, if we wish to excite electrons from the valence band to the conduction band, we need to supply only $(1/2)E_g$. This energy is normally called **Fermi energy (E_F)** of electrons of an intrinsic semiconductor.

Can we alter this energy? Yes, this can be altered, if the intrinsic semiconductor is doped with atoms possessing valence either greater or smaller than that of the host atoms. For example, the energy of valence electrons of silicon (which is four) can be altered by adding either boron (which has three valences) or phosphorus (which has five valences). Again, an explanation of the reasons for the alteration of energy of electrons requires various concepts. Therefore, without going into much detail, we can assume that the Fermi energy of an electron is related to the concentration of dopant (like phosphorus in silicon) present in the semiconductor. The addition of phosphorus, for example in silicon, makes it n-type, because each phosphorus atom creates one additional free electron in silicon. It is possible to increase the energy of valence electrons of silicon by such doping so much that its energy is almost equal to the lowest energy level of the conduction band. Under this condition, pure silicon which is an insulator can become a conductor like metal without altering its semiconducting properties. However, for practical purposes, the concentration of doping is maintained such that the difference between the Fermi energy of the electron and lowest energy level of the conduction band (E_c) is about 0.1–0.2 eV (Fig. 7.1c, d).

7.3.3 n- and p-Type Materials

If silicon is doped with boron, then each boron atom creates a scarcity of one electron (i.e., increases concentration of positive charge known as **hole**) at the site where boron is substituted. This type of semiconductor thus has a large concentration of holes and such semiconductor is referred to as a *p*-**type semiconductor**. Therefore, unlike doping silicon with a phosphorus atom, each boron atom decreases the concentration of the electron by one. Since Fermi energy is related to the number of electrons present in the system, and since Fermi energy of intrinsic silicon is equal to $(1/2)E_g$, any decrease in electrons due to boron doping will decrease Fermi energy from its intrinsic Fermi level. A decrease in the Fermi level is shown by shifting its position from the intrinsic level toward the valence band. In other words, Fermi energy will shift toward the valence band as we increase the concentration of boron atoms in silicon. In order to distinguish the shifting of Fermi energy due to boron from phosphorus, the former is referred to as $_pE_F$. For practical purposes, like phosphorus, doping with boron is done such that the difference between the Fermi energy $(_pE_F)$ and the uppermost level of valence band (E_V) is maintained at about 0.1 eV (Fig. 7.1e, f).

Thus, we see that silicon can be doped either to make *n*-type with its Fermi energy almost near the conduction band (Fig. 7.1c), or it can be doped to make *p*-type with its Fermi energy almost close to its valence band value (Fig. 7.1d). In other words, while *n*-type material can be viewed as material having excess of electrons, *p*-type material can be viewed as having a very low concentration of free electrons (or having a large concentration of holes). Thus, if an *n*-type material is brought in contact with a *p*-type material, there would be a natural flow of excess electrons from *n*-type to *p*-type, until an equilibrium has been established.

7.4 Formation of *p* : *p* Junction

What is the effect of flow of electrons on joining *n*- and *p*-type materials? It is necessary to realize that if there is an electron transfer from one material to another, the material losing electrons becomes positively charged and the other becomes negatively charged. When we join *n*- and *p*-type materials, and if electron transfer occurs in the fashion expressed earlier, then both materials become electrically charged. This behavior is very difficult to digest, as materials cannot become electrically charged by simply joining them! In reality, electron transfer does occur, but instead of an electron leaving its parent material, it accumulates at its interface. As a result, the neutrality of the material is maintained and yet electron transfer occurs.

7.4.1 Formation of Space Charge Region

Since, in solid material, atoms are rigidly fixed in a given lattice configuration, any movement of electrons automatically creates a positive charge at the lattice site from where the electron has moved out. Therefore, it is not unreasonable to assume that while electrons accumulate at the interface of two semiconductors, an equivalent amount of positive charge would be created within *n*-semiconductor. It can be confirmed mathematically that these disturbances due to the transfer of electrons take place within the vicinity of about 10,000 Å from the interface of the semiconductor. It is assumed that a positive charge created within the n-semiconductor also lies in a plane at about 10,000 Å from the interface. Distance between this plane and the interface is known as the width of **space charge** (w). The magnitude of "w" depends on the difference between the Fermi levels of two semiconductors; larger the difference, greater is the width of space charge. This model suggests the formation of two types of charges present in two planes separated by the width of space charge. Moreover, these planes behave like two parallel plate capacitors. Let's assume for simplicity that the potential difference created due to such charge separation in a semiconductor is about 1.0 V and the charges are separated by a distance of 1000 Å. Then, an electrical field of 0.1 MV cm^{-1} is created in the space charge width. Electrical field of this magnitude will be formed in both n- and p-type Si semiconductors.

Under this situation, if we excite an electron from the valence band to the conductance band in the n-semiconductor, a hole will be created in the valence band. If these carriers (i.e., electrons and holes) are created within the space charge region, electrons will be immediately forced to collect at the other end of n-type material (i.e., behind the interface) and holes would be forced to move to the interface of n- and p-semiconductors. Similarly, if an electron of p-semiconductor is excited within the space charge region, due to electrical field, the electron will move toward the interface of n- and p-semiconductors and hole will move toward the bulk. If excitation of electrons is done in both semiconductors simultaneously, electrons will find its place at the backside of the n-semiconductor, and holes will be collected at the backside of the p-semiconductor. If the backside of the n- and p-type materials is connected to an ammeter, these accumulated carriers will move to get neutralized, resulting in a flow of current.

7.4.2 Distribution of Carrier Concentration

Another aspect of this junction formation needs to be touched upon. Before n- and p-type materials are joined, the concentration of electrons (in n-type) and holes (in p-type) are uniformly distributed throughout the material. In other words, there is no accumulation of these carriers at any specific place in the material.

However, when these materials are joined, some of the electrons get accumulated at the interface (for reasons explained earlier) in n-type and holes in the p-type. If we wish to represent the variation in concentration of these carriers in the material, it would be reasonable to express their concentration in an exponential fashion. In other words, after the junction formation, the concentration of electrons appears to decrease exponentially as we go from the interface to the bulk of the material. Similarly, the concentration of holes also decreases exponentially as we go from the interface to the bulk of p-type material. Conceptually, a hole is a representation of missing electrons, therefore, it may not be incorrect to represent the concentration of holes as missing of electron. Pictorially, formation of $p - n$ junction and variation in

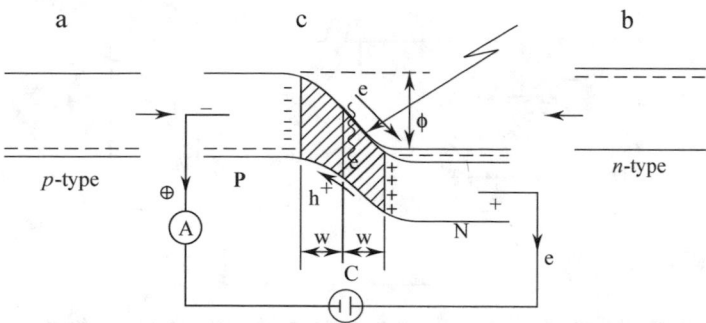

Fig. 7.2 A schematic diagram of $p-n$ junction formation. **a** and **b** show the band diagram of p- and n-type materials before the junction is formed. **c** shows the condition of these two materials after the junction is formed. The shaded portion is space charge width of n- and p-type materials. ϕ is the magnitude representing the difference between the Fermi levels of the two materials. This value is also known as contact potential. The back ends of both materials are connected to an ammeter and a small DC power source. When some radiation falls in the space charge width, the newly created electrons and holes follow the direction as shown by the arrow, which are collected at the back end of the material to form a current

concentrations of these two carriers in the two materials are shown in Fig. 7.2. This is an oversimplified view of the junction formation.

7.4.3 $p-n$ Junction vis-a-vis Diode

What is the advantage of forming such junction? Advantage of the $p-n$ junction is that it can allow the flow of electrons only from p- to n-type materials and not vice versa. The direction of flow of current is decided by the direction of the electric field formed in each semiconductor. Therefore, this type of junction acts like a **rectifier**. Moreover, if $p-n$ junction is short-circuited, there would almost be no flow of current. But if the p-type material of $p-n$ junction (Fig. 7.3) is externally biased by a positive potential, electrons tend to flow in the circuit. An exponential relationship is observed between the magnitude of the applied potential and the current flowing through the circuit. This is given as

$$I = I_0 \exp\left(\frac{eV_0}{KT} - 1\right) \cong I_0 \exp\left(\frac{eV_0}{KT}\right) \tag{7.1}$$

where, I, I_0, and V_0 are current flowing at applied potential, and (V_0) and I_0 are the current flowing in the absence of any applied potential. K and T are the Boltzmann constant and absolute temperature, respectively. This suggests that when potential is increased, current increases exponentially. However, when the p-semiconductor is applied a negative potential, the above equation takes a different shape as given here

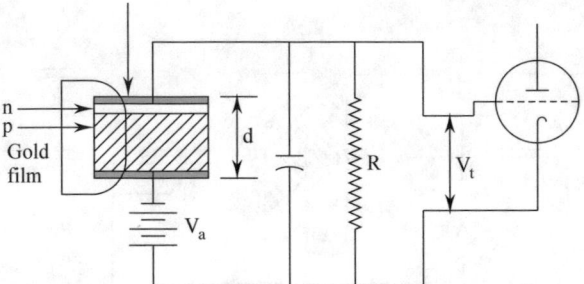

Fig. 7.3 Schematic diagram of a $p-n$ junction. $p-n$ junction is formed with p-type material (shown by slanted lines) and n-type material. The gold film is deposited on the back of both semiconductors for making an ohmic-type contact. Radiation falls from the back of the n-type material. V_a is the small DC battery. V_t is the negative pulse measured by a diode. R and C are the resistance and capacitor to prevent any leakage of current

$$I = I_0 \exp\left(1 - \frac{eV_0}{KT}\right) \cong I_0. \tag{7.2}$$

Thus, I would be independent of negatively applied potential and would be almost equal to I_0. This suggests that when the p-type semiconductor is connected with a negative terminal of the potential, and n-type with positive terminal, then current would become independent of the applied potential. These two equations suggest that when an AC current is allowed to flow through a $p-n$ junction, only positive current will appear as an output. In other words, a $p-n$ junction behaves like a rectifier. Now we shall discuss its application in radioactivity measurement.

7.5 Effect of Radiation on the $p-n$ Junction

If any radiation like γ-ray, α- or β-particulate radiation is allowed to interact with a $p-n$ junction, and if the energy of these radiations is greater than the band gap, it can excite electrons from the valence band to the conduction band, thereby creating holes in the valence band. These electron–hole pairs experience an electric field present in the space charge region and hence separate instantaneously. The electrical field of the space charge region forces electrons to flow toward n-type and holes toward p-type. If the two extreme ends of the $p-n$ junction are connected to an ammeter, current flows in the circuit. In this manner, electrons from n-type eventually come back to p-type through the external circuit to get annihilated. Magnitude of current is proportional to the quantity of radiation interacted with the $p-n$ junction.

When excited electrons and holes from the space charge region arrive at the back of the n-type and p-type semiconductors, respectively, a potential is developed and the magnitude of potential corresponds to the energy of the radiation interacted with the semiconductor. Thus, a measurement of potential developed after the $p-n$ junction was exposed to these radiations gives information about the energy of interacting radiation, provided the potential developed across the $p-n$ junction is calibrated in terms of energy (MeV). In other words, one single radiation interacting with $p-n$ junction produces one pair of electron/hole. The number of such electron/hole pairs produced is equal to the number of potential pulses produced. Therefore, measurement of the number of such potential pulses corresponds to the number of radiations interacting with the $p-n$ junction. Therefore, with the help of a pulse height analyzer and an amplifier, one can identify and measure the activity of the sample.

This description is an oversimplified version of what actually happens at the $p-n$ junction after illuminating it with such electromagnetic radiation. In reality, we do not measure the potential, which is developed across the junction. In fact, the junction is formed due to the difference in the Fermi levels of the two semiconductors. This produces a contact potential (ϕ volt). The generation of electron/hole pairs decreases the magnitude of this contact potential. Hence, as in ionization counters, we actually measure the decrease in contact potential (like decrease in anode potential of a proportional counter) when the radiations interact at the $p-n$ junction. As ionization counters, this decrease in the contact potential appears as a negative pulse.

The efficiency of generation of electron/hole pairs by radiation largely depends on the feasibility of their absorption within the space charge width. Therefore, thickness of the $p-n$ junction, especially the side through which radiation has to interact with the junction, should be kept to a minimum value. Thin material allows the radiation to penetrate through the material to reach the space charge region. Thickness of the semiconductor which is exposed to the radiation should, therefore, be kept to a few microns, so that even weak β-particle or a particle can penetrate to reach the space charge region of the $p-n$ junction (Fig. 7.3).

7.5.1 Design of a Semiconductor Detector

Considering these factors, a typical design of a $p-n$ junction and its electronic circuit are shown in Fig. 7.3. This circuit resembles a proportional counting system. The $p-n$ junction (Fig. 7.3) is made of p-type (shown by slanting lines), on which the n-type semiconductor is kept such that there is no air gap in between. There are many techniques to achieve this configuration. In order to collect electrons (from n-type) and holes from the back of p-type, a thin layer of conducting material, normally gold, is deposited over both semiconductors. But one of the main considerations is that the Fermi energy of the metal must match with the Fermi energy of the n- and p-type materials, so that one can achieve an ohmic contact and not a junction-type contact. The entire thickness of the $p-n$ junction (d) is kept to approximately 0.5 cm or less. The entire system is sealed in a metal enclosure, with n-type being left open for the radiation to penetrate. The p-type is connected to the positive terminal of a battery (normally 2.0 V DC battery). A capacitor and a resistance are joined in parallel to avoid counting of any spurious pulses. The negative pulses are initially fed through a diode for measurement, followed by feeding through an amplifier, pulse

Table 7.1 Energy of radiation required to create electron/hole pairs with some materials. The energy of radiation to be counted should be greater than band gap

Material	Energy required for ionization/eV
Silicon	3.23
Germanium	2.24
Indium antimonate	0.06
Cadmium sulfide	5.2
Gray tin	0.1

height analyzer, and then to the counter, in a similar fashion as was done with a proportional counter.

7.5.2   Advantage of a Semiconductor Detector

Semiconductor detectors have a very thin window due to the thin film of gold and the thin layer of n-type (in the present case, but it can be p-type as well for the $n : p$ junction). Therefore, even weak β-particles, soft X-rays or α-particles can be counted with high accuracy. Since the detector needs only about 1.5 V as its operating voltage, there would hardly be any spurious pulses produced, as normally observed in scintillation counting systems or proportional counters, i.e., noise-to-signal ratio is very low. Since the size of the detector is very small (about $3 \times 2\,cm^2$), its effective area of the detector is very small. Hence, background activity due to cosmic radiation is also very low. For α-counting, shielding is not required as pulses for α-particles are much larger than that produced by cosmic radiation. The resolution of β-peak is also very good. 4π geometry counting can also be easily achieved with this type of counter without making the counting assembly bulky. Moreover, dead time for such detector is of the order of 10^{-7}–10^{-8} s, indicating that one can count radioactive materials with a high activity using this detector. Some common types of semiconductor detectors are given in Table 7.1.

7.6   Silicon- and Germanium Lithium-Drifted Detectors

There are some specialized types of semiconductor detectors, which always need to be kept under liquid nitrogen temperature; most popular among them are silicon doped lithium and germanium doped lithium detectors. The most impressive application of these detectors now appears to be in the field of very low energy X-ray spectroscopy (0.2–20 KeV). Using a 5 mm diameter and 3 mm thick Si (Li), it is possible to detect the K-line of carbon (997 eV). The window thickness (1 mm Ge) limits the use of Ge (Li) for measuring very low energy X-ray, while for energies above 20–30 KeV, higher efficiency of measurement makes them preferable to a Si (Li) detector.

7.7 Nuclear Emulsion Techniques

In radiochemical work, sometimes it is necessary to separate the constituents of the mixture before carrying out any radioactive measurements. This is normally done using the paper chromatography technique. In such paper chromatography technique, it is also necessary to establish the position of various components of the radioactive mixture which has moved over the chromatography paper. Sometimes it is also necessary to measure a cumulative radiation dose received by the personnel working in a radiochemical laboratory. In all such cases, it becomes difficult to measure radioactivity by the techniques discussed previously. For these types of specific requirements, nuclear emulsion techniques and other related techniques are normally adopted. We shall now concentrate on these techniques in some detail.

7.7.1 Silver Grains in Photographic Plate by α-Particles

When charged particles pass through a photographic emulsion, they produce latent images along their path. Upon development of the film, a grain of silver appears along the tracks of the particles. A lot of valuable information can be obtained from the study of the tracks. Counting individual particles along the paths gives a measure of the number of nuclear particles entering the plate. A study of the detailed structure of the track leads to the determination of mass, charge, and energy of the particle. Film sensitivities are controlled primarily with sensitizers.

The energy of α-particles causing blackening of the photographic plate can be calculated by measuring accurately the range of each track. The energy loss by α-particle can be calculated by counting the grain present in each track, because the rate of energy loss is proportional to the number of grains developed per unit path length.

In addition, the sensitivity of a given type of film can be varied within wide limits by techniques used for developing the track. The development of film, however, needs special care so that tracks are not lost or twisted during development. The shrinkage of the film during drying is avoided by soaking the film in glycerin, which fills up all the voids, which are left by the removal of the unsensitized silver halides.

7.7.2   X-ray Film Badges

The photographic film is widely used to monitor the dose of X-ray, γ-rays, or β-particular radiation received by the personnel. For this purpose, a new unexposed film is worn by a worker and developed to measure the blackening density after a period of time. On comparison of the blackening density with a calibrated curve, the amount of dose received by the person can be deduced. For calibration purpose, a photographic film (X-ray film) is exposed to various amounts of known radiation for a definite time (to calculate the dose rate). The exposed film is developed and the density of the blackening of the film is measured. A graph is plotted between the density of the blackening of film and the amount of dose given to the film. This graph is used as a calibration graph to find out the amount of dose received by a person wearing the film.

Film badges are covered with a light protector as well as with small metallic plates of different thickness made of different materials. Material is selected such that it stops radiation entering the film to a different degree depending upon the nature of radiation passing through these plates (Fig. 7.4A, B, C, D). One portion of the film contains only a light protector but no metal (Fig. 7.4F). After developing the X-ray film, a study of the relative darkness produced in each square of the film suggests the nature and the quantity of radiation received by the person wearing the film badge. The film badge shows total accumulated radiation received by the person over a period for which the film badge was worn. The disadvantage of this method is that the amount of radioactivity received by the person is known after the film is developed, which is usually after about a fortnight of its use. This is because the film badges are sent to the concerned authority for development after either every fortnight or a month. Nevertheless, for the accumulative dose, this is a standard method followed these days.

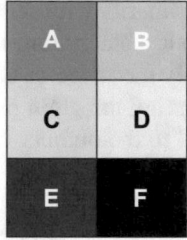

Fig. 7.4 A schematic figure of a film badge showing the arrangement of different metal plates used for monitoring different types of radiations received by the person wearing the film badge. A, B, C, D, and E are small metal pieces made of different metals of different thicknesses to differentiate the type of radiation received by the person bearing the badge, and the area F gives total exposure, because it does not have a metal piece

7.7.3 Emulsion Radio Chromatography

An emulsion technique usually known as "**Autoradiography**" is used in paper chromatography or paper electrophoresis, to locate the separated constituents from a mixture of radioactive substances and to measure the relative intensity of radiation present at each constituent. This is achieved by the following procedures.

7.7.3.1 Experimental Procedure for Autoradiography

In this technique, the solution which is to be analyzed or separated is placed on a sheet of a long filter paper (about 6 in. or longer) in the form of a small spot near one end of the paper (leaving about 1 in. space from the bottom of the paper). Then the paper is dried slowly and hanged in a glass jar containing a suitable solvent. The height of the paper is adjusted such that the solvent touches the bottom of the paper only. Care has to be taken to ensure that the paper does not move. The solvent is allowed to pass lengthwise through the paper. Various components of the original mixture move with the solvent, but at different speeds. After the solvent has moved a sufficient distance, the paper is dried (e.g., by an infrared lamp). In order to recognize the position of the original spot and the solvent front, a fresh spot of the radioactive mixture is put at the original spot, and a line is drawn with the radioactive mixture to mark the solvent front. For illustrating this procedure, the following experiment was carried out. A solution containing NaBr and KBr was irradiated in a thermal nuclear reactor. This mixture was separated by paper chromatography by adopting the procedure as discussed previously. A solution containing HCl in methanol (1 : 1 v/v) was used as a solvent to separate the mixture. Location of the constituents (including the solvent front and position of the original spot) were found by using any of the methods discussed here:

1. The paper chromatogram was counted by scanning a G.M. counter and count rate was recorded over a strip chart recorder (Fig. 7.5). This graph shows the positions of radioactive materials separated at different places on the paper chromatogram.
2. The paper chromatogram was exposed overnight on an X-ray film. This film was developed and printed (Fig. 7.6a). The photograph was used to calculate the R_f factor, as positions of original spot and solvent front were visible in the print.
3. In another experiment, 11 layers of X-ray films were kept one over other on the paper chromatogram. This was then exposed overnight. The top (i.e., 11th layer) X-ray film (Ilford industrial G) was developed and its print was taken (Fig. 7.6b). This photograph gives an idea about the relative penetrating power of the radiations emitted by radioactive nuclides present in the mixture. It can be seen that except for β-radiation emitted by ^{24}Na, all other radiations emitted by other isotopes have been able to penetrate 11 layers of the X-ray films.

The R_f values obtained from paper chromatogram by G.M. counter and from X-ray films as well as the energy of radiations emitted by these isotopes are shown in Table 7.2.

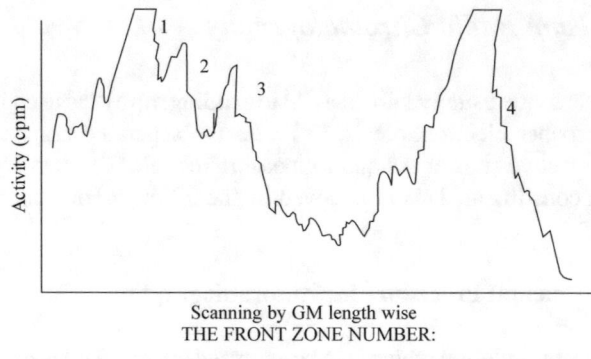

Scanning by GM length wise
THE FRONT ZONE NUMBER:

1. ORIGINAL SPOT   2. POTASSIUM IONS (^{42}K$^+$)
3. SODIUM IONS (^{24}NA$^+$)   4. BROMIDE IONS (^{80}Br$^-$)

Fig. 7.5 A plot of count rate versus paper length of a paper chromatogram used for separating a mixture of radioactivity present as ^{24}Na,^{80}Br, and ^{42}K. The paper chromatogram was scanned while keeping G.M. counter fixed at one position. Peak of activity 1 = original spot on the paper, 2 = confirmed to be ^{4}2K$^+$ ions, 3 = confirmed to be ^{24}Na$^+$ ions, and 4 = confirmed to be ^{80}Br-ions

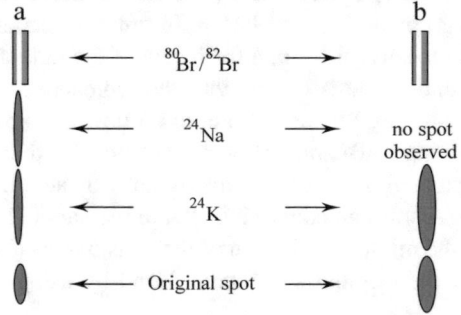

Fig. 7.6 Autoradiograph of a paper chromatogram showing the separation of different constitutes of a nuclear reactor irradiated mixture of NaBr and KBr. **a** Photograph of X-ray film which was kept over the paper chromatogram for exposure in dark. Spots are identified by their half-life measurements **b** Photograph of X-ray film which was kept over the paper chromatogram for exposure; in between the film and photograph paper, 11 layers of X-ray films were kept and then exposed in dark. The spot of ^{24}Na has disappeared because its radiation could not penetrate 11 X-ray films

Table 7.2 Relative E_{max} (in MeV) for β-particles emitted by these nuclides, the observed R_f factors obtained by chromatogram scan and X-ray films

Nuclide	E_{max} (MeV)	R_f factors	
		G.M. scanning	X-ray film
Potassium-42	3.60	0.16	0.16
Sodium-24	1.39	0.29	0.24
Bromine-80	1.99	0.81	0.83

It is customary, however, to calculate the R_f factor for each separated nuclide; this is defined as

$$R_f = \frac{\text{distance traveled by the spot-front}}{\text{distance traveled by the solvent-front}}.$$

The R_f factor of an ion has a characteristic value under a given condition, and may give a preliminary indication of the ion's identity. Relative $E_{\max}$ (in MeV) for β-particles emitted by these nuclides, the observed R_f factors obtained by chromatogram scan, and X-ray films are given in Table 7.2, to show the closeness of the two experimental results.

R_f factor of certain ions may be small or very close to the values of other species present in the mixture (like ^{42}K, in this example); this makes a satisfactory separation difficult. In such cases, a multiple development technique may be helpful. A developed strip (as described earlier) is dried and re-chromatographed, (after cutting the strip of the paper chromatogram containing those spots which have been separated satisfactorily, e.g., spots corresponding to ^{80}Br and ^{24}Na and leaving behind the spot which needs to be separated further, e.g., ^{42}K) in the same direction with the same solvent. This grants more time to spots of low mobility or almost similar R_f values to separate. Alternatively, a zone containing two or more unseparated substances may be cut from the chromatogram, sewed to a second strip, and re-chromatographed with a different solvent. However, this method suffers from the disadvantage of requiring a longer time, and might not be useful for short-lived radioactive isotopes.

7.7.3.2 Precautions Needed

One needs to take care that the amount of radioactive sample should not overload the paper as this leads to a tailing effect and bad separation of the nuclides. On the other hand, carrier-free samples may not move so readily on the paper. Hence, proper precautions are needed to get good separation. Selection of the paper is also important, and some factors such as presence of impurities, wet strength, thickness, and flow rate are important factors to be taken into consideration while carrying out paper chromatography separation. Paper chromatogram should not be dried by blowing with air, because this may cause some of the radioactive nuclide to blow away from the paper. The paper must be properly dried especially if the isotope is a weak β-emitter, otherwise, the residual solvent absorbs some of the radiation. This is important when the paper is going to be counted directly by the G.M. counter. Alternatively, each separated constituent can be extracted from the paper, and then it can be counted by any one of the methods discussed in the previous chapters. Many radioactive nuclides produced by a nuclear reaction have short half life. In such cases, it must be separated and identified rapidly. Paper chromatography can be carried out rather rapidly by choosing a fast-moving solvent and short paper strip. Usually less viscous liquid will move faster, and descending chromatography gives a faster flow rate than the ascending one.

7.8 Solid-State Track Detector

In recent years, solid-state track detectors are gaining ever-increasing importance as a powerful experimental tool in the study of a number of nuclear phenomena. These detectors are made of insulating materials such as mineral crystals (e.g., quartz glass, and mica sheet), glass (e.g., pyrex and sodalime), and certain synthetic plastics (like lexan, makrofol, etc.). The system takes advantage of the fact that energetic heavy ions such as fission fragments moving through the insulating material leave behind narrow continuous trails of damage which can be made visible under an optical microscope after suitable chemical etching. Chemical etching is based on the principle that large free energy associated with the disordered structure makes damage trails more chemically reactive than the normal material. If a substance containing damage trails is immersed in a suitable chemical, those damaged sites which intersect a surface are preferentially leached.

The technique has several advantages over conventional semiconductor counter or the time-consuming radiochemical methods of fission studies, the most attractive being its simplicity. The detector is very selective and effectively discriminates against processes other than fission and thus affords convenient means for studying fission events, in an essentially background free situation.

7.9 Low-Level Counting

In low-level counting (for isotopes other than β-emitters), especially for measuring food contamination by radioactive isotopes, on fall out of nuclear explosion to the atmosphere, etc., one is not concerned with the efficiency of counting, but in reducing the background activity to a value as low as 0.2 cpm. Because in low- level counting the activity of a sample may be of the order of 10–30 cpm. With a background of ordinary G.M. counter (whose background counts are normally 10–20 cpm), it becomes impossible statistically to distinguish activity due to either the sample or the background. One can think of reducing the background by increasing shielding of the counter with a thick lead wall. But one cannot reduce the background to less than 10–20 cpm even by making lead of 30 cm thickness. Lead, being the end product of all radioactive series, is bound to be contaminated with a few natural radioactive substances, which can contribute toward the background radiation. Hence some modification to the assembly of the G.M. counter and some modification to the electronic circuit are made to lower the background to almost 0.1 cpm. This is achieved by two processes namely **Anti-Coincidence counting** and **Co-incidence counting** systems. These are discussed here.

7.9.1 Anti-coincidence Counting System

By an anti-coincidence counting system, background activity can be reduced to 0.1 cpm. This is usually done with a thin mica end-window G.M. counter, because other counters need a large space and hence would be bulkier. A typical design of this counter is shown in Fig. 7.7. Two G.M. counters A and B are used, with B having a large aluminum window and a small-sized counter A (mica-type window). The source to be counted is kept very close to counter A such that its radiation does not pass through both counters simultaneously.

However, cosmic radiation or γ-radiation (emitted by isotope present in the lead shield) can pass through both A and B counters simultaneously producing negative pulses in both counters at the same time. The electronic arrangement is made, such that if negative pulses are produced in both counters simultaneously, they are not allowed to pass through the counting system (i.e., scaler) and are rejected, but if produced in any one of the counters, they are allowed to enter the counting system (i.e., to the scaler). Naturally, it will always be the inner counter, which will receive radiations by the sample, unless it could penetrate the wall of the inner counter to reach the second counter, which is difficult to encounter under the present geometrical arrangement. Rejection of negative pulses or its acceptance is achieved with the help of an anti-coincidence electronic circuit.

With a shielding of 3.5 cm of mercury and 20 cm of steel, it has been possible to reduce the background to 0.77 cpm, or shielding of 3 cm of lead and 10 cm of steel could give a background to about 1.3–1.5 cpm. These types of counting systems

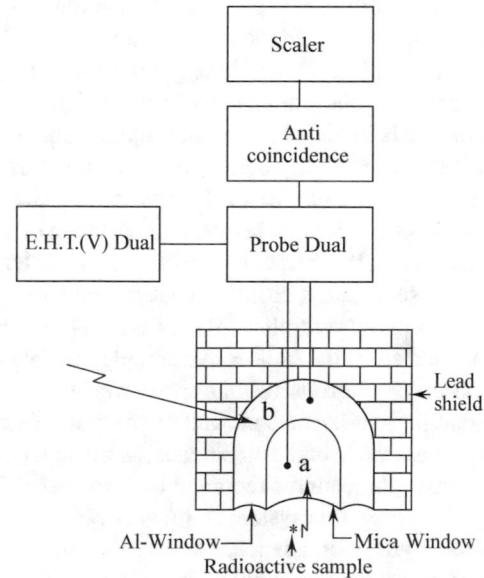

Fig. 7.7 A schematic diagram of an anti-coincidence counting system with two G.M. counters under the anti-coincidence condition. A is a mica window G.M. counter and B is a large aluminum end-window G.M. counter

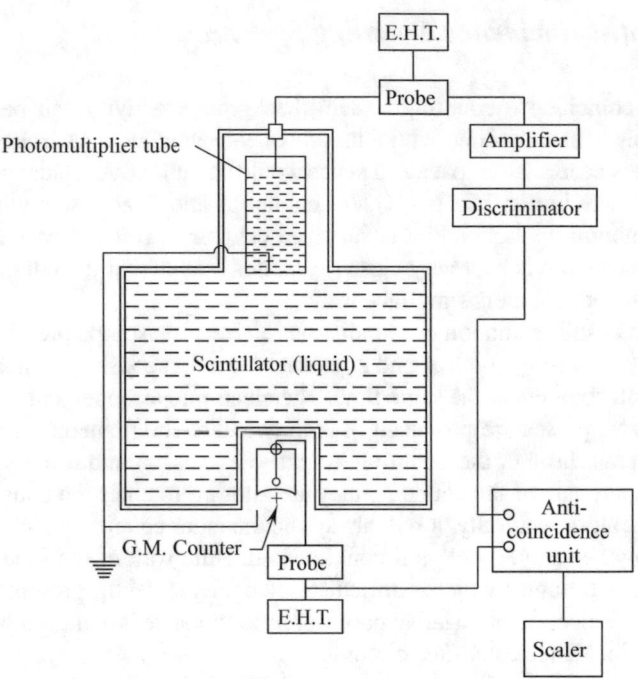

Fig. 7.8 A schematic diagram of an anti-coincidence low background counter using a photomultiplier tube and a G.M. counter in anti-coincidence to each other

are not useful for α-particulate counting because of the problem of penetration of α-particles through the mica window of the G.M. counter.

Alternatively, a low background β-counting unit can also be set up with a G.M. counter in anti-coincidence with a liquid scintillation counter. The scintillation counter is made of a photomultiplier tube and a vessel containing the liquid scintillator, which can count β-particles more effectively than γ-radiation. The vessel is constructed of two steel tubes fitted concentrically. In order to get a sufficiently good optical contact between the cathode of the photomultiplier tube and the liquid scintillator, the former is immersed in the latter (Fig. 7.8). Because of the large area of liquid scintillator, cosmic radiation and strong β-particulate radiations are observed with a photomultiplier tube as well as the end window G.M. counter. β-particles emitted from the radioactive sample are detected by an end-window G.M. counter. Cosmic background radiations as well as strong β-radiations are detected by a liquid scintillation detector as well as the small G.M. counter simultaneously. These two counters are in anti-coincidence, as a result, when radiation is seen by a smaller G.M. counter, the count is recorded to the scaler.

However, this system is not very popular compared to the use of two G.M. counters in anti-coincidence, because the latter has no problem of liquid handling and protecting the photomultiplier tube from light.

7.9.2 Co-incidence Counting System

This type of counting does not decrease the background, but can decrease noise level produced in the counter due to the heating of some electronic components or dynode of a photomultiplier tube in the scintillation counting system. Usually, this unit is used with a scintillation counter for measuring the activity of low energy β-particulate radiation.

The principle of this unit is based on the fact that the noise level in the two photo-multiplier tubes or in any electronic components of the instrument is not produced at the same time. In this type of counting system, generally two photomultiplier tubes are separately connected with EHT and amplifier unit. The output of both tubes is fed to a coincidence unit, which is connected to a pulse height analyzer followed by a scaler. The sample to be counted is put in between the two photomultiplier tubes. If photons are seen by both the photomultiplier tubes, the coincidence unit allows the pulses generated by these tubes to be analyzed by the pulse height analyzer, else it is rejected. This way noise produced in the photomultiplier tubes or in amplifier units is eliminated from being recorded. The elimination of noise level is essential for counting samples like ^{14}C and ^{3}H. The efficiency of ^{14}C counting can be achieved as high as 80% and that of ^{3}H 50% by such coincidence counting system.

Summary

In this chapter, we learnt some special techniques and detection systems for measuring radioactivity for special purposes. These techniques would be useful when dealing with studies of some nuclear reactions and identification of the products formed in a very small number. From a public health protection point of view, we have to ascertain that milk, food, or air are not contaminated with radioactive materials either due to some accidents that occurred in nuclear thermal power plants, or due to some nuclear explosion for military applications. For such purposes, we need a counter which can measure activity in the range of a few fractions of a microcurie. We saw the type of counters available for this purpose. In many nuclear reactions, we face the problem of analyzing α-particles. This type of counting can be done effectively with solid-state detectors. We also learn about the separation of the radioactive isotopes from a mixture of radioactivity present in a sample by the radio-chromatography technique and the techniques to evaluate the efficiency of the separation technique.

Chapter 8
Sample Preparation for Counting

8.1 Sample Preparation Techniques

Radioactive nuclides separated by any radiochemical techniques will probably need to be counted. The separated material cannot be simply put near the counter and counted. A special technique is required to prepare a suitable source for counting. In this chapter, various modes under which a radioactive material can be counted and importance of source preparation for counting activity of the radioactive sample is presented.

8.1.1 Counting of Solutions

Counting of radioactive materials by liquid scintillation method is difficult when they are not soluble or miscible with the liquid scintillator. In some cases, insoluble substance is forced to dissolve, by addition of a lipoidal base or acid to form toluene soluble salt. Some metallic cations are converted to toluene soluble complexes by treatment with a suitable chelating agent. When several sources are to be compared, care is needed to ensure that quenching is either absent or maintained constant in all samples. However, liquid sample for liquid G.M. counter does not need any special preparation, except corrections for density, volume of liquid, etc.

8.1.2 Counting in Form of Suspension

Sometimes, either it is not desirable to dissolve a material in any solvent or not possible to dissolve in a suitable solvent (solvent which does not cause quenching in scintillation counter). In such cases, the following technique can be adopted. A suspension of fine powder of a material is made by means of some suitable gelling agent. Two types of gelling agents: Cab-o-Sil and thixine are commonly used, as they require no heating for forming gel. The gelling agent is blended into the scintillation liquid and an aliquot is poured into a counting vial containing fine powder sample. The vial is shaken to disperse powder evenly. The vial may need slight heating to form a gel. This gel holds powder in suspension. This method ensures thorough mixing of the radioactive substance with the scintillator. The vial with gel mixture is counted for measuring its activity.

8.1.3 Counting by Spreading Sample over a Filter Paper

This is comparatively a recent technique, which offers high promise for substances which are insoluble in toluene or other liquid scintillator bases. Either thinly spread sample obtained by filtration through a Millipore filter paper or liquid sample soaked filter paper is dried and placed directly into the counting vial. Liquid scintillator is poured through the side of the vial without disturbing the sample. The vial is counted by placing it over a photomultiplier tube. Filter paper on addition of liquid scintillator becomes transparent to visible light, hence scintillations formed can be detected by the photomultiplier tube.

Paper counting technique is also excellent for counting radioactive material separated over the paper-chromatograms. For this purpose, paper-chromatorgram is cut into equal segments and each segment is inserted into the counting vial to which liquid scintillator is added and then counted. Since the radioactive spot is insoluble in the scintillator, the paper can be removed from vial and radioactive material can be extracted from segment if required.

8.1.4 Deposition of Sample by Electrolysis

Radioactive samples for β- and α-counting are prepared frequently by the electrodeposition technique. This technique gives uniform and thin sample. As a result, losses of activities due to self-absorption are avoided. This technique gives excellent result with small quantity of radioactive nuclide, specially when it is too low to prepare thin source by precipitation method.

The advantage of this technique is usage of simple apparatus and high yields in short time. Furthermore, stable sources of high quality can be obtained. Electrodeposition is done on a thick electrode so that source remains flat during transfer for counting. Moreover, care has to be taken such that the deposition occurs only on one side of the electrode. This is achieved by insulating all sides of the electrode, except for one side, including the wire that is used for soldering the electrode for making the connection. The efficiency of electrodeposition can be determined by measuring the ratio of activities deposited on the electrode and that remaining in the aliquot. This method, for instance, has been applied for preparation of sources used in determination of the half-life of some uranium isotopes. The electrolytic deposition can be done with or without the use of carriers. For example, a very thin film of Silver-110 can be obtained by deposition of Silver-110 on a gold cathode from a slightly ammoniacal solution of silver argentocyanide labeled with Silver-110. This gives about 97.98% of Silver-110 deposition. This deposition can be made with or without the presence of natural silver as a carrier. Source prepared without the carrier would suffer negligible loss due to self-absorption.

8.1.5 Vacuum Evaporation Technique

Using vacuum evaporation technique, the source can be prepared by volatilizing by heating under reduced pressure.

The sample gets deposited over the substrate. Sometimes difficulties arise as the evaporated layer becomes thicker at the substrate, leading to pulling of the sample. Thus, source quality is determined by quality of backing on which the thin source has been prepared. If plastic foil is used as a substrate, it must be cooled by placing a thick copper plate in contact with the substrate. Size and area of the deposit can be controlled by placing a mask of the required size over the substrate during evaporation process. Although this technique gives a thin film of a material, it increases the decontamination problem, because during vacuum evaporation in the vacuum chamber, the material gets deposited on the glass jar and other places. Before selecting this technique, the above problem must be taken into consideration.

8.1.6 Electrospraying Technique

Electrospraying technique can also be used for thin source preparation. This technique makes use of the fact that when a potential is applied between a capillary containing a polar liquid and a plate at ground potential, the liquid is ejected from the capillary as a fine spray. A solvent with a sufficiently high vapor pressure is chosen so that it gets evaporated before reaching the ground plate. The dissolved material then gets deposited as a thin layer. In this case, as well, the quality of source is partly influenced by the quality of substrate. Generally, a positive high tension of 5–10 kV is applied to the capillary.

8.2 Solid Sample Source

We learnt some special methods for sample preparation, but these are not commonly used in radiochemical work. Most of the time we prepare solid sample source for measuring the activity. Now, we shall devote our discussion on solid sample preparation.

Sample preparation is an important parameter to get an accurate activity of the sample, specially when the activities of series of samples are to be compared from each other. In radiochemical experiments, the radioactive sample can be either in liquid, gas, or solid form. The sample in gas form can be counted directly by injecting it into gas chamber of the counter (e.g., G.M. or proportional counter). Liquid sample can be directly counted by a liquid G.M. counter or liquid scintillation counter (i.e., by using either an organic scintillator, or an organic phosphor or well type NaI scintillator).

Solid sample, however, needs special attention, especially for α-source and weak β-source. γ-rays have strong penetrating power, thus sample preparation does not have much influence on efficiency of counting. However, selection of method, for source preparation, largely depends upon the nature and energy of radiation of the sample to be counted. For α-source and β-source, a thin homogeneous and stable source is needed to prevent loss of activity due to self-absorption. This is important if specific activity of the sample of α-emitter or low energy β-emitter is to be determined. To prepare a source for solid counting, of weak β-source or α-source, it is necessary to make a thin and even source. Moreover, when activity present in a series of samples are to be compared, thickness of each sample must be the same and should be uniformly even throughout the source (Fig. 8.1A and not like Fig. 8.1B, C, or D).

The sources, except for shown in Fig. 8.1A, cause loss of radiation due to self-absorption of varied degree due to the variation in thickness of the source (Fig. 8.1B, C, and D). Thus, the count rate for samples B, C, or D may lead to a wrong speculation.

In addition to making a sample of uniform thickness, magnitude of thickness of the sample also plays an important role, especially when comparisons of activities of various samples are to be made.

The following example would be able to explain the effect of variation in thickness on the count rate. In this experiment, variation in the source thickness is been achieved by adding various amount of $BaCl_2$ solution to a set of different test tubes containing same amount of $Na^{35}SO_4$ solution, $BaCl_2$ present in the test tube gets precipitated by adding $NaSO_4$ solution which results, in giving $Ba^{35}SO_4$ precipitate, with different amount of $BaSO_4$, but same amount of radioactive ^{35}S. The precipitate from each test tube, after several washings with water, is transferred to source tray forming an even surface (Fig. 8.1A) of $BaSO_4$ labeled with Sulfur-35. Since each sample contains same amount of ^{35}S, one would expect the count rate of each sample to be the same. Contrary, one initially observes a decrease in count rate with increase in amount of $BaSO_4$ (Fig. 8.1E) after some amount of $BaSO_4$ the count rate becomes independent of the amount of $BaSO_4$. This thickness of the material is called as **saturation thickness**. This effect can be hypothetically explained by dividing the thickness of $BaSO_4$ by infinitesimally small thin layers. β-particle of each of these infinitesimally thin layers will try to reach the counting system. As the number of layer increases, absorption of β-particles starts to take place; some of them are absorbed by the layers while others escape the layer to reach the counting system. Thus, it is possible to imagine certain fixed number of layers of $BaSO_4$ which will always allow a fixed number of β-particles to reach the counting system. A thickness lesser than

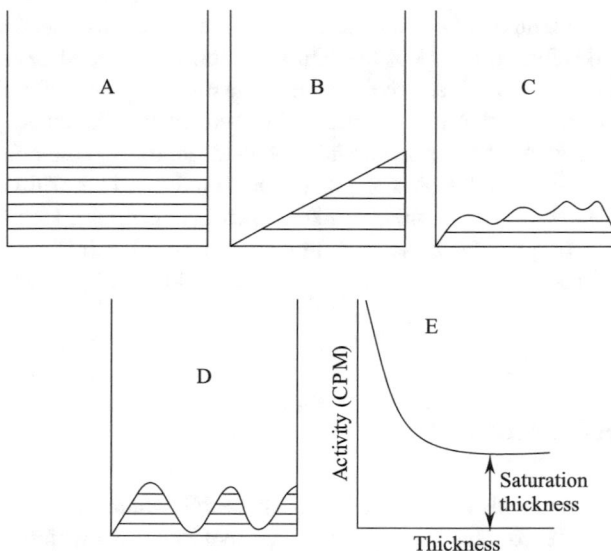

Fig. 8.1 A schematic diagram showing the possible variation in the thickness of samples: **A** Sample prepared with uniform thickness, **B** Sample with slanting thickness, **C** Non-uniform sample thickness, **D** Samples with many hills, and **E** a graph showing variation in count rate of same amount of activity versus thickness of the sample showing a saturation thickness

this number of layers will allow more number of β-particles to reach the counting system, whereas a greater thickness will allow the same number of β-particles to escape the sample to reach the counting system. Therefore, though the count rate measured at the saturation thickness is less, it becomes independent of thickness of the sample (provided thickness is uniform like shown in Fig. 8.1A). Hence, when one needs to study variation of activity in different solid samples, it is desirable to find the saturation thickness and make all samples of thickness equal to the saturation thickness. Thus, if there is any variation in activities from such set of samples, one can be sure that it is not due to variation in the thickness of sample but due to experimental conditions only.

The saturation thickness, obviously depends upon nature and energy of radiation being measured and atomic weight of the solid material being precipitated. As a rough guess, normally the saturation thickness would be equal to E_{max} of β-particles or E_a of α-particles. One can, thus, easily estimate approximate thickness of the sample needed to achieve the saturation thickness. It goes beyond saying that such precautions in making the counting samples are necessary with only with low energy β-particles (i.e., E_{max} smaller than about 0.5 MeV) or α-particles of any energy. Alternatively, one can make a very thin source evenly prepared, which is capable of absorbing a very small fraction of radiation due to self-absorption, especially for weak β-sources or α-sources. But many a time it may not be possible to restrict ourselves to small amount of material for the source preparation; one may have to use a thicker source or one may have different amounts of samples to be counted to study the relative change in the radioactivity of the sample. Under these conditions, one should experimentally check the saturation thickness, as discussed earlier. After having obtained information about the saturation thickness, in all experiments one should add enough non-radioactive material to the counting sample, before making the counting source, so that all samples to be counted has thickness greater than the saturation thickness. Under this counting condition, though the count rate measured at the saturation thickness is less, it becomes independent of thickness of the sample (provided thickness is uniform like shown in Fig. 8.1A). Hence, when one needs to study variation of activity in different solid samples, it is desirable to find the saturation thickness and make all samples of thickness equal to the saturation thickness.

8.2.1   Planchet Material

Solid sample source needs to be prepared on a material, which can be easily handled and placed near the counter for counting. The material onto which the source is prepared is called planchet. This is made of a circular shape to match the shape and size of the counter's window. Thickness of the material is kept such that it does not twist easily and remains sturdy and flat. The type of material used for the planchet is also important. This is normally made of aluminum, copper, glass, plastic, etc. However, one has to take care that metal of the planchet does not react with the solvent (being used for drying the materials).

Acids may etch aluminum, stainless steel, or copper. For such purpose, we either use acid resistant metal. If a metal like aluminum is to be used, then such planchet is regarded as dispensable. Glass or plastic planchet may sometimes be used for acidic solution, or even platinum for some purpose. In order to keep the backscattering uniform, planchet made of same material should be used for all experiments, especially when comparison of activities of a series of samples is desired.

8.2.2   Preparation of Solid Sample from Liquid Sample

Sometimes we may like to evaporate the solvent to get a solid substance for counting, especially when the amount of radioactive material present in the sample is very low to be precipitated. However, while drying a solution of sample it gives rise to non-homogeneous solid sample source. Sample gets concentrated in form of big crystals at the boundary of the original drop, resulting in a rather high and non-reproducible sample, which gives rise to varied degree of self-absorption (Fig. 8.1D). This is avoided by making surface of the sample holder (planchet) hydrophilic by addition of wetting agent prior to dispersing the drop of radioactive solution. Care is taken, however, to add minimum amount of wetting agent in order to avoid unnecessary addition of solids (i.e., wetting agent) to the source. Addition of artificial centers of crystallization to the radioactive liquid also improves the source quality. These artificial centers help the solids (i.e., radioactive materials) to crystallize in many small crystals instead of a few big ones. Big crystals cause loss of activity due to uneven spread over surface of the sample holder.

8.2.3   Source from a Slurry

In most of the radiochemical work, we come across with a problem of counting freshly prepared precipitate. Following procedure is adopted for making a sample source for counting purpose. From the solution of radioactive material, we precipitate the desired chemical, which is filtered and washed several times with suitable solvent. Precipitate is transferred to a test tube and a slurry is made in some suitable solvent (e.g., water, acetone, methanol, etc.). Slurry is transferred slowly with a micro-pipette to the planchet and then it is dried under the infrared lamp. Care is taken to get uniform deposit of the precipitate over the planchet (Fig. 8.1A). The planchet along with the dried solid is then measured for its activity, by a suitable technique. In order to prevent loss of activity by splashing while drying the sample, the source is dried in a gentle air stream at about 40 °C and is protected from dirt. Crystals of the solid sample can be stabilized by adding a few drops of binding agent, prior to drying (e.g., 0.5% collodin in acetone or 2% isobutylmethy-lacrylate polymer in ethylacetate). Addition of a trace of secptine to the slurry prior to drying prevents the dried source to not become powdery. A trace of detergent, such as teepol, often assists preparation of even deposits. Cracking of the solid sources while drying can often be prevented by addition of a few drops of celite, which is very useful when drying gelatinous precipitates.

Alternatively, the slurry can be transferred to a filter paper fitted in a micro-funnel; this is convenient when the solid material has to be washed thoroughly. The filter paper is not allowed to dry until the entire process of transferring the slurry and washing, etc., has been completed. The filter paper along with the precipitate is transferred to a planchet coated with thin layer of wax. Waxing the inside surface of the planchet allows the filter paper to remain flat while slowly drying under the infrared lamp. This method is most successful because self-absorption can be made very negligible.

Summary

The discussions made in this chapter revealed that source preparation is one of the important parameter to get a reproducible result, especially when series of samples are to be counted or when specific activity of the sample is to be determined. In this chapter, we have learnt a detailed account of various methods normally used for the preparation of sample source. In the next chapter, we shall discuss the factors other than sample preparation, affecting the efficiency of counting.

Chapter 9
Factors Affecting the Counting Efficiency

9.1 Introduction

In the previous chapter, we realized the importance of source preparation for getting meaningful results. However, in addition to source preparation, there are many other factors, like geometrical factors, scattering of radiation, selection of radiation emitted by the isotope etc., that also need to be considered for measuring the activity of the radioactive sample. It is necessary to consider these factors, especially when an absolute count rate is required. In this chapter, we shall discuss these factors to a greater depth.

The best way to determine counting efficiency is by counting a sample whose absolute count rate is known, e.g., x disintegration per minute (abbreviated as x dpm). Suppliers of the isotope give this information. If the observed count rate of the isotope (containing x dpm) is y cpm, then the efficiency of counting system is given by the following equation:

$$\left.\begin{array}{r}\text{Efficiency}\\ \text{factor}\end{array}\right\} = \frac{\text{Excepted count rate } (x \text{ dpm}) - \text{Observed count rate } (y \text{ cpm})}{\text{Excepted count rate } (x \text{ dpm})} \quad (9.1)$$

This procedure is most accurate, simple, and takes care of all possible errors which might be present in the selected counting procedure and which perhaps may not be covered in the factors discussed in this chapter. In the foregoing sections, we shall discuss some of the factors, which can affect the efficiency of the counting process.

9.2 Geometrical Efficiency

In counting system, where the source is mounted outside the sensitive volume of the counter, such as in the end-window G.M. counter, the efficiency of counting depends upon the fraction of radiations that pass through the window of the counter. The

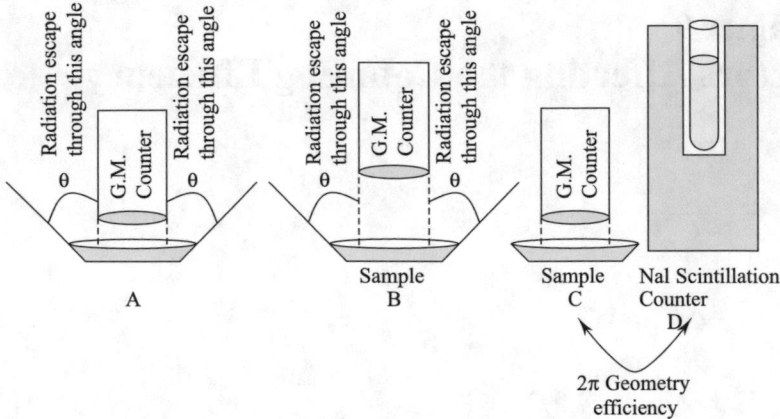

Fig. 9.1 Schematic diagrams showing (**A**) the arrangement of end-window G.M. counter placed such that solid angle created by the sample is larger than the diameter of the window of the counter, (**B**) that in spite of the fact that the diameter of the window and sample matches, positioning of the counter is such that radiation escapes the window of the counter (**C**) the positioning of window such that diameter of the window and its position with respect to sample allows no radiation to escape, (**D**) the liquid sample being counted in a NaI well-type counter. Both arrangements as shown in C and D give 2π geometry

source has a finite diameter, and the way to achieve maximum solid angle at the window is to mount it as near as possible to the counter (Fig. 9.1). If the distribution of the source on the tray is uniform and symmetrical (Fig. 8.1A), then it is possible to calculate the geometrical efficiency as a function of the vertical distance from the window and the horizontal displacement from the central line of the symmetry. The source diameter should be made less than the diameter of the counter.

When a solid is mounted inside the counter e.g., in proportional counter or scintillation counter (Fig. 9.1D), the geometrical efficiency is approximately 2π.

In the scintillation counter, the shape of the phosphor can influence the geometrical efficiency of the system. On the contrary, in gas-flow type proportional counter or well-type NaI(Tl) crystal (for scintillation counter), the sample is kept inside the sensitive volume of the counter (Fig. 9.1D), hence geometrical efficiency will be as high as 2π (i.e., radiation emitted by the source in the solid angle of 180° can be counted by the counter). When the sample (over the planchet) is kept outside the counter, the planchet makes a solid angle with the window of the counter. In order to cover the entire solid angle with the window (so that the maximum percentage of radiation emitted by the source could reach the window of the counter), the planchet needs to be kept as near as possible to the window (Fig. 9.1C) of the counter (but certainly not so close that it touches the window, because then it will contaminate the window with the radioactive substance).

With the scintillation counter using plastic phosphor, the shape of phosphor can influence the geometrical efficiency of the system. For example, a well-type NaI(Tl) crystal can ensure that a high proportion from γ-ray emitted by the sample passes through the crystal, or a plastic phosphor may be machined to such a shape that it offers the maximum surface area to the solution of β-emitter.

Reproducible geometrical conditions are often achieved more readily by counting liquids rather than solids (as it is very difficult to prepare an identical solid sample source).

9.2.1 Self-absorption

In Chap. 8, we discussed that when sources for counting purposes were prepared from the solution containing different amount of $BaSO_4$ with the same amount of ^{35}S activity, a saturation thickness was observed. In this experiment, the activity was found to decrease with an increase in the amount of $BaSO_4$. Alternatively, we can repeat this experiment in a slightly different fashion. In a test tube, let us add a fixed amount of $BaCl_2$ to a solution of Na_2SO_4 labeled with ^{35}S. Then the entire amount of $BaCl_2$ is precipitated as $BaSO_4$. The test tube is centrifuged and the supernatant is decanted. The precipitate is washed several times with water. A slurry of the precipitate is made. Ten planchets are taken and are numbered as 1 to 10. Starting from planchet No. 1, the slurry is added to each planchet in increasing amount i.e., 1, 2, 3, 4, 5, 6, 7, 8, 9, and 10 ml up to the 10th planchet. During the addition, care has to be taken that the slurry should not overspill. For this purpose, planchets are kept under the infra-red lamp and as the solution dries and the required amount of slurry is added slowly. This way we have 10 planchets containing an increasing amount of $BaSO_4$ labeled with ^{35}S. After the planchets are dried, the activity of each planchet is calculated using a G.M. counter and the calculated activity versus amount of $BaSO_4$ is plotted. Care is taken that the precipitate has formed a uniform thickness (like Fig. 8.1A). One would expect to observe a linear increase in the activity with the amount of $BaSO_4$ added, but on the contrary, the activity increases exponentially with an increase in the amount of $BaSO_4$ (or so to say with an increase in the thickness of the source) and reaches to a maximum value (constant value) beyond which there is no increase in activity in spite of the fact that the planchet contains more activity of $BaSO_4$ labeled with ^{35}S (Fig. 9.2). The thickness corresponding to this maximum activity is known as a **saturation thickness**. Any further increase in the source thickness beyond this value does not increase the count rate further.

Why do we observe such behavior? In Chap. 8, we explained the reasons for getting this behavior. But it is important to realize that in the earlier experiment, the total activity of Sulfur-35 added to each planchet was the same, but the amount of $BaSO_4$ precipitated in each planchet was different. As a result, the saturation thickness activity was minimum. In the present experiment, the activity added to each planchet with the increase in total weight of $BaSO_4$. In other words, though activity, as well as the weight of the precipitate increased in this experiment, but yet we get the effect of saturation thickness. In order to understand this behavior, let's divide the entire thickness of the precipitate into many imaginary infinitesimal thin layers of $BaSO_4$

Fig. 9.2 A graph showing the variation in the activity of BaSO$_4$ labeled with Sulfur-35 versus amount of BaSO$_4$ labeled with Sulfur-35 added to the precipitate

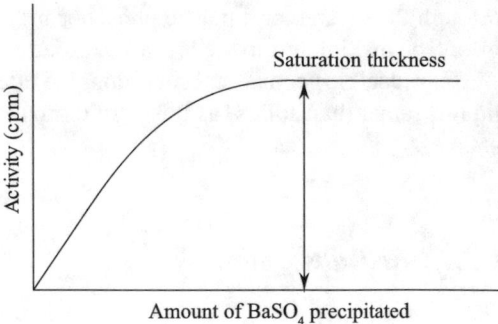

and examine the effect of an increase in the number of layers on probability escape β-particles from the bottom of each layer. β-particles try to escape the precipitate to reach the window of the counter. When the number of layer is first or few from the bottom, then the probability for the β-particles to penetrate the layer is more. Therefore, β-particles are able to escape the layers, with an increase in the number of the layers (or increase in the amount of BaSO$_4$). This will cause an increase in activity measured by the counter. But as we keep on increasing the number of layers of BaSO$_4$, a time comes when the total thickness of the precipitate (let's call it x number of layers) becomes equivalent to maximum thickness β-particle (from the bottom of the layer). Any further increase in the number of layers will be equivalent to adding that many number of layers on the top of the maximum number of layers (i.e., x number of layers) and decreasing the same number of layers from the bottom of the x number of layers, the net result is that though we have added activity to the planchet, the net activity reaching the window of the counter remains the same. Thickness lower than this x number of layers allows more β-particles to reach the window of the counter. Hence, as we increase the amount of BaSO$_4$, in the beginning, the activity increases as well, but as soon as we reach this minimum thickness required to stop all β-particles emitted from the bottom of the first layer, the activity becomes independent of the thickness. As a result, any further increase in thickness has no effect on the increase in the activity. This phenomenon is due to the process known as **self-absorption** of β-particles by the thickness of the sample itself.

The effect of self-absorption is appreciable with β-particles of low penetrating power e.g., Sulfur-35, Nickel-63 etc. In counting of solid samples, to keep the loss of radiation due to self-absorption constant, the uniformity of the source distribution over the source tray is very important (Fig. 8.1A). In all comparative measurements, either weight of the sources should be constant, or minimum thickness (saturation thickness) should be calculated, so that, loss in activity due to self-absorption is minimum in all the samples. This correction is important, especially when large number of samples is to be counted and variation in activity is expected to occur. This correction factor can also be evaluated by a theoretical method. But it is much easy to maintain the source thickness of all samples greater than the saturation thickness, by adding some non-radioactive substance prior to precipitation. When the liquid is counted, self-absorption remains constant as long as densities of the liquids are not changed.

9.2.2 Backscattering

An important manifestation of the scattering of β-particle is the increase in count rate obtained when a source is mounted on a solid backing material. The additional count rate is due to β-particles which are scattered toward the counter through an angle of 180°. This effect can be explained by measuring the activity of a radioactive sample kept over a planchet with an end-window G.M. counter. The sample is kept near the window of the counter and beneath the planchet, a shelve is made to insert some metal sheet. First, we measure the activity of the sample. And then we measure the activity of the sample by inserting a thin aluminum sheet into the shelve. This process is repeated by inserting an aluminum sheet of increasing thickness and the corresponding increase in the activity is measured. One would expect that since we have not changed the activity of the sample, or its position from the window, the count rate in all these measurements should approximately be the same. On the contrary, activity keeps on increasing exponentially as we increase the thickness of the aluminum sheet, reaching a maximum value for some particular thickness of aluminum. Any further increase in the thickness, does not lead to an increase in the activity. The nature of the graph is similar to one represented in Fig. 9.2.

Why should activity increase when nothing has been done to the radioactive sample nor more radioactive material has been added? The radioactive sample emits the radiation in all direction (i.e., in 4π direction). Those radiations moving toward the window are recorded in a normal fashion (i.e., within 180°). But those radiations which are emitted at angle 270° or near about (i.e., $> 180° < 360°$), are reflected toward the window by the aluminum sheet kept beneath the planchet (Fig. 9.3). As a result of this reflection, the activity recorded in the counter increases. As we increase the thickness of the aluminum sheet, more and more radiations are reflected toward the window. But this process of reflection comes to a maximum value when further increase in thickness does not allow radiations to get reflected, but some of them get absorbed by the aluminum sheet, hence are not in a position to reach the window of the counter. Thus, a saturation value reaches after a certain thickness of the aluminum sheet. This phenomenon is known as **backscattering**. Theoretically, the limiting value for increase in count rate is reached when the material is thick enough to absorb all backscattered radiations from the deepest layer and are unable to penetrate the absorber to reach the window of the counter. The backscattering phenomena increases with an atomic number of the element used. Hence, to minimize the errors caused due to this effect, the source tray (planchet) is made of a material with a low atomic number. In any comparative work, if the same source tray and the source is used, this effect can be neglected.

Fig. 9.3 A schematic
diagram to explain the
process of the back reflection
from the aluminum sheets

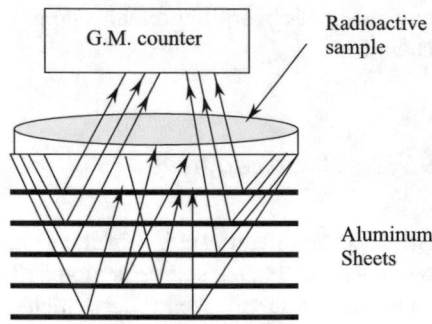

9.2.3 External Absorption

Absorption of radiation occurs when it passes through a medium. The magnitude of this absorption depends on energy and nature of radiation (i.e., α-, β-particles or γ-rays) and nature of the medium into which the radiation enters (i.e., air, density of liquid, or atomic weight of the solid). In an end-window G.M. counter, normally the source is kept very close to the window. In other words, some amount of air gap is kept between the source and the window. Thus, the radiation emitted by the source has to pass through the source thickness (this aspect has been considered in the previous section), air thickness (due to air gap between the source and the window), a sheet covering the source (like polythene sheet), and window thickness, before it can reach the active zone of the counter for being recorded. Absorption of radiations by these substances has an effect on the efficiency of counting. These factors are very important, especially in counting low energy particulate radiations.

The thickness of the window can be minimized by using a thin sheet of mica. Mica thickness can be made as thin as 2 mgcm^{-2}, whereas aluminum window has a thickness of about 30 mgcm^{-2}. Thickness of air can be minimized by keeping the source close to the window. However, the thickness of 1 cm air (atmospheric) is about 1.3 mgcm^{-2}. Hence, the absorption is negligible in the G.M. counter. If the source is covered by cellulose (often used to cover the source), then its thickness is about 10 mgcm^{-2}. This covering should be avoided as far as possible. Therefore, while counting a sample of week β-particles, thickness due to these materials must be taken into consideration. Theoretically, one can also calculate the % loss of activity due to this thicknesses. This calculation can thus help in selecting the type of counter to be used for radiation measurement. In the liquid G.M. counter, radiation has to penetrate the glass wall of the counter. Thickness of the glass wall is of the order of 20–30 mgcm^{-2}. Hence, the efficiency of counting for low energy of β-particle is low with a liquid G.M. counter. Since these factors affect radiation loss due to absorption of radiation by materials other than the source itself, the loss of radiation is called as loss due to **external absorption**. In order to prevent losses of activity due to this aspect, a minimum of material should be kept between the window and the source.

Nevertheless, these effects are not important in case of gas-flow proportional counter as it does not have a window nor air gap between source and counter.

9.2.4 Background Activity

Results of source counting are always associated with the natural background. Background count is subtracted from observed count rate, to obtain the count rate due to the source only. In G.M. counter, the background count rate is usually a constant value and is not very high (between 10–20 cpm), whereas, with a scintillation counter, the background count rate depends much on the operating conditions and is usually high (with exception of α-counting). In the proportional counter, background count also depends on the type of particulate radiation being counted. For example, background with α-particle counting is low, as the discriminator may be set to reject pulses produced by cosmic radiation and not the large α-pulses. However, when β-radiation is counted, usually the background count is high.

9.3 Decay Scheme

In measuring the activity of a radioactive isotope, information regarding the nature of radiation and its energy is very useful in selecting a counter. In addition, if we examine various radioactive isotopes, it will be observed that almost all of them emit more than one type of radiation with a different mode of emission. Information regarding the type of radiation being emitted, its percentage abundance, energy of radiation emitted, half-life etc., are shown in a decay scheme of each isotope (Figs. 9.4–9.8). The percentage of decay mode by a particular path is helpful in selecting a counter and the nature of source (i.e., whether in liquid or solid form) for counting an isotope.

For example, Cobalt-60 isotope (Fig. 9.4) emits β-particles as well as γ-rays. Energy of γ-rays is 1.17 MeV while that of β-particles 0.31 MeV. γ-rays being more energetic, one would not prefer to measure the activity of cobalt by counting low energetic β-particles. Moreover, low energetic β-particle would need lot of care in sample preparation. In addition, a simple end-window G.M. counter can

Fig. 9.4 Decay scheme of Cobalt-60 isotope

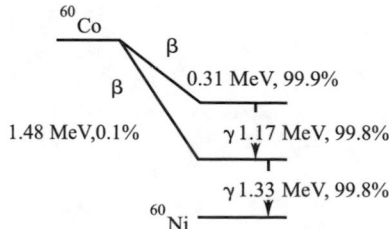

be used for counting energetic γ-rays. Therefore, if we have to measure the activity of Cobalt-60 isotope, we should prefer to count its γ-rays rather than β-particles. Its decay scheme also reveals that although Cobalt-60 decays by strong β-particles (1.48 MeV), its abundance is only 0.1%. This indicates that out of 100 atoms of Cobalt-60, only 1 atom is likely to decay giving β-particle of 1.48 MeV. In other words, most atoms of **Cobalt-60** will prefer to decay by β-particles of energy 0.31 MeV followed by the decay of γ-rays of energy 1.17 MeV. Decay scheme also indicates that this isotope decays by β-particles (0.31 MeV) and γ-rays (1.17 MeV) with a similar percentage. Since it is easier to count γ-rays, this isotope can be counted either in liquid or solid form by a G.M. counter. Thus, we see that a knowledge of the decay scheme helps in selecting a counter for counting its activity. It can also help in deciding the form in which the isotope should be counted in order to get maximum counting efficiency. It can also give information regarding possible interference of radiations, emitted by the radioactive materials on the actual count rate. To clarify these points further, few decay schemes are discussed here.

9.3.1   Tritium

Tritium decays by a low β-particle having an energy of 0.185 MeV (Fig. 9.5). Its half-life is very long (12.35 years). This isotope cannot be counted by the end-window counter, since the energy of β-particle is low and radiation is not able to penetrate the thickness of the window. It is, therefore, counted by gas-flow proportional counter (in solid form) or scintillation counter, especially with coincidence counting (in liquid or solid form). In case of liquid sample e.g., aqueous solution or colored organic substance labeled with tritium, quenching correction is required. Since half-life is very long, counting can be done for a longer period to minimize the error of counting. All these decisions can be taken by simply examining the decay scheme of tritium.

Fig. 9.5 Decay scheme of tritium

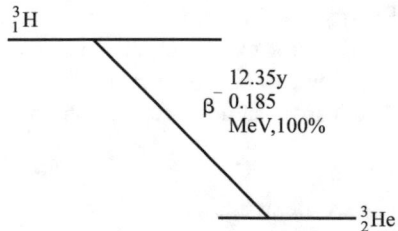

Fig. 9.6 Decay scheme of
Sodium-22

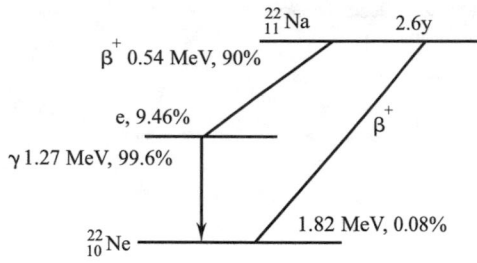

9.3.2 Sodium-22

Sodium-22 decays with a half-life of 2.6 years (Fig. 9.6) and 90% of its decay is
by emission of β^+-particle (0.54 MeV) followed by γ-emission (1.27 MeV). In
another mode, this isotope decays by emission of one β^+-particle (1.82 MeV) with
only 0.08% abundance and another β^+-particle (0.54 MeV) is emitted with 90%
abundance which is followed by electron capture decay with an abundance of 9.46%.
All these processes finally end by γ-rays emission of 1.27 MeV energy.

Therefore, although the decay of Sodium-22 takes place by emission of β^+ with
high energy, its counting efficiency is low due to either its low abundance (0.08%) or
it gets annihilated with electrons producing radiations with balanced energy. On the
other hand, it has γ-rays of 1.27 MeV with an abundance of 90.0%. Therefore, one
should try to count γ-rays of Sodium-22 isotope with NaI(Tl) scintillation counting
method. This isotope, therefore, can be counted either in liquid or solid form. However, in γ-spectrometry, one would also be able to observe an annihilation peak of
β^+ at 0.54 MeV.

9.3.3 Sodium-24

The half-life of Sodium-24 is only 15 h. Hence, this isotope can be used only for
experiments where the maximum time required for the completion of experiment is
not more than 20–30 h so that experiment and counting could be finished within one
or two half-lives.

Sodium-24 decays (Fig. 9.7) by strong β-particles (1.39 MeV) with 99.9% abundance. It is associated with γ-rays of two energies 2.75 MeV and 1.36 MeV with
abundance around 99.9%. γ-rays with 2.75 MeV energy while interacting with
matter can also produce a pair production of 1.02 MeV, which after annihilation
would produce two photons of 0.511 MeV. Considering these factors, it is clear
that this isotope should be detected by measuring either β-particles (1.39 MeV)
by any counter or γ-ray of 1.36 MeV by NaI(Tl) scintillation counter. It should
be noted that, in γ-spectrometry, photo peaks corresponding to energies 1.02 MeV,
0.511 MeV, and 2.75 MeV due to pair production, annihilation, and its own γ-
rays, respectively, can be observed. Since the efficiency of the production of pho-

Fig. 9.7 Decay scheme of
Sodium-24

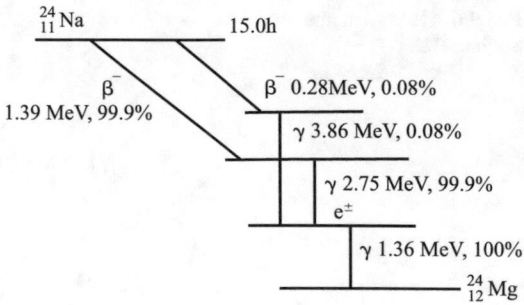

toelectric effect would be more with γ-rays of lower energy, intensity of photo-peaks will appear in the order of 0.511 MeV > 1.02 MeV > 2.75 MeV. Due to the high energies of β-particles or γ-rays, the sample can be counted either in liquid or solid form and unless very high accuracy of counting is required, liquid or end-window G.M. counter can also be used for counting this isotope. Moreover, while counting by G.M. counter, it would be difficult to isolate count rates due to γ-rays (2.75 MeV) and β-particles (1.39 MeV). Therefore, though it would be easier to detect its activity by a G.M. counter, it would be difficult to calculate absolute activity due to interferences of these two radiations. Because these radiations get added up in counting, thus making the observed count rate higher than the expected from the decay of one nucleus of Sodium-24.

9.3.4 Strontium-90

Strontium-90 is a pure β-emitter (Fig. 9.8) with very long half-life (28.5 years). Energy of β-particle is low (0.546 MeV). Hence, care is necessary for source preparation. It can be counted by the end-window G.M. counter or scintillation counter. However, ^{90}Sr undergoes a secular equilibrium with ^{90}Y. The equilibrium is established within 2–3 d. After the establishment of secular equilibrium activity of Strontium-90, it becomes equal to that of Yttrium-90. Hence, instead of measuring the activity of

Fig. 9.8 Decay scheme of
Strontium-90

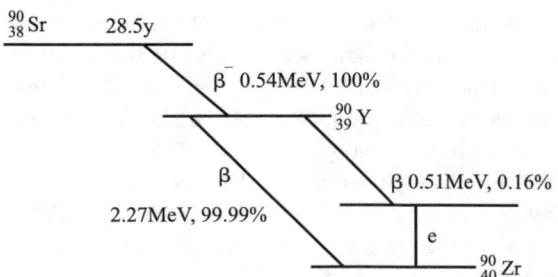

Fig. 9.9 Decay scheme of
Cesium-137

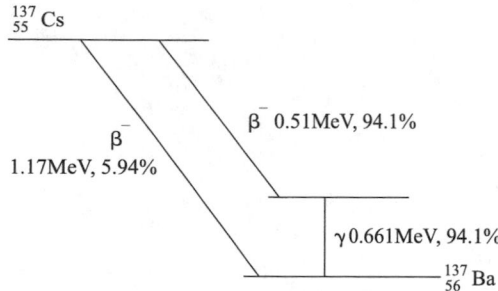

^{90}Sr, one prefers to count the activity of ^{90}Y. Because, ^{90}Y has a strong β-particle (2.27 MeV) with 99.9% abundance. Therefore, the counting efficiency for counting Yettrium-90 is higher. This can be counted by an end-window or liquid G.M. counter. Source preparation would also require less attention to make a uniform sample. Considering these advantages, Sr^{90} isotope is preferred to be counted after allowing the sample to achieve secular equilibrium. For this purpose, one carry a normal experiment with Strontium-90 isotope, which is counted after 3 d.

9.3.5   Cesium-137

Cesium-137 isotope has a very long half-life (Fig. 9.9) but gives low energy β-particles with high abundance and γ-rays with 94% abundance. The efficiency of counting, however, would certainly be higher with γ-rays, than β-particle (0.51 MeV), unless plastic phosphor scintillator or proportional counter is used. However, if high efficiency is not essential, an ordinary G.M. counter would be sufficient for this isotope.

Summary

In this chapter, we have learnt the importance of source preparation and loses of activities due to several factors like self-absorption, backscattering etc. What precautions are needed to minimize the effect of these factors are also discussed. Finally, utility of knowing the decay scheme of radioactive isotopes at least for making a decision as to in which form the sample should be counted, which radiation to be used, and by which counter the isotopes should be counted is illustrated by discussing the decay schemes of few radioactive isotopes of long- lived and short-lived type.

Chapter 10
Identification of Radioactive Isotopes

10.1 Introduction

In radiochemical work, our interests are to identify the isotope, as well as measure the activity accurately present in the sample. Identification of isotope is necessary for analytical work like activation analysis, nuclear reaction studies (where nature of isotope formed is unknown), type of radioactive isotope present in contaminated food etc.

In most of these cases, one can identify the isotope by determining the half-life of the isotope, as well as by the nature of radiations emitted by the isotope (by obtaining its spectrum or determining its E_{max}). Half-life measurement and the nature of radiation emitted by the isotope also helps to establish the purity of the isotope present in the sample (i.e., whether the sample contains several radioactive isotopes or a single type isotope). This information assists in selecting a suitable radiochemical analytical technique to separate isotopes present in the sample, before its activity is evaluated. The next step is to examine the decay scheme of the identified isotope. Decay scheme helps in deciding the type of counter, nature of sample (liquid or solid), type of radiation to be counted to measure the activity present in the sample.

10.2 Counter Selection

From the above mentioned preliminary investigations, we are able to decide whether a particular isotope can be counted by a G.M. counter (i.e., for strong β-particles or γ-rays), or proportional counter or scintillation counter. Nevertheless, since our objective is to identify the isotope, we can resort to either proportional counter or scintillation counter. Using either of these two counters, the energy of radiation, as well as exact nature of radiation, can be confirmed. This is achieved by studying the energy spectrum emitted by the source with the help of a pulse-height analyzer. If isotope is confirmed to emit α-particles, for example, then either a semiconductor detector or a gas-flow proportional counter or a liquid phosphor scintillation counter is preferred for measuring its activity.

10.3 Energy Determination

It is useful to find out the energy of radiation emitted by the isotope. β-particles fortunately can be classified by evaluating their maximum energy known as E_{max}. For this purpose, one normally uses a scintillation counter to get the entire spectrum or a G.M. counter to get the absorption spectrum. α-particles or γ-rays, being monoenergetic radiations, their energies can be found out by measuring their spectrum. These techniques are discussed here.

10.4 β-Spectrometry

While discussing applications of scintillation and proportional counting systems in the previous chapters, it was pointed out that with the help of a pulse-height analyzer, the spectrum of β-particles can be obtained. Since β-particles possess energy ranging from zero to a maximum value known as E_{max}, one is normally interested in knowing the value of E_{max} rather than the entire spectrum of β-particles. E_{max} value gives an idea in regards to penetration power of β-particles. This is a very useful information for deciding the method of counting and for protecting human body from the health hazard point of view. Though it is difficult, to get an idea of E_{max} from the β-spectrum, nevertheless it is preferred to measure this value from the absorption curve of β-particles.

10.4.1 β-Absorption Law and Its Spectrum

The absorption of β-particles closely follows the law of exponential nature i.e.,

$$I = I_0 e^{-\mu x} \tag{10.1}$$

where I_0 is initial number of β-particles reaching one end of the absorber (i.e., when $x = 0$), I is number of β-particles penetrated out of the absorber (i.e., for thickness

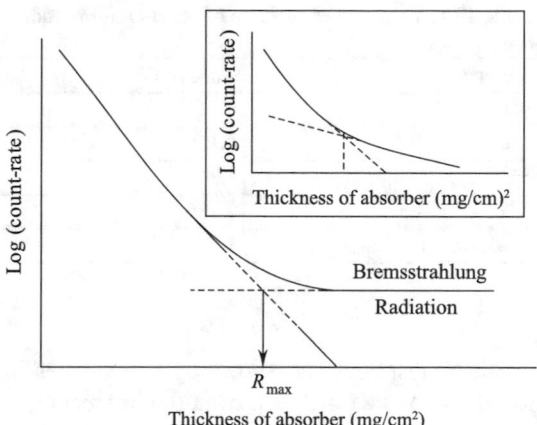

Fig. 10.1 A schematic representation of a absorption curve for a β-particle showing the R_{max} of the β-particles

of the absorber x cm) and μ is the linear absorption coefficient (cm^{-1}). Therefore, when log of (I) is plotted against the thickness of the absorber, a linear relationship is obtained. The thickness at which the count rate becomes zero is considered as its **maximum thickness** which is equivalent to the required thickness of the absorber to stop all the β-particles passing through it. A detailed procedure for this purpose is discussed here.

One of the standard techniques employed to obtain information about maximum energy of β-particle is by plotting an absorption curve. For this purpose, a β-particle source is kept near the window of a G.M. counter, keeping some gap between the source and the window. This gap is required to insert the absorber in between the source and the window of the counter. The count rate is measured by varying the thicknesses of an absorber (normally aluminum metal thin sheet). Prior to keeping the source near the window, background activity is measured. This measurement is taken in the absence of the source and absorber. This count rate is subtracted from each measurement obtained when counting was done in the presence of the absorber. By this process, we get the actual number of β-particles (count rate) that penetrated the absorber. In order to get the exact value of thickness equivalent to E_{max}, count rate is plotted versus thickness of the absorber (aluminum sheet) on a semi-log graph paper. Alternatively, the log of count rate can be plotted versus the thickness of the absorber on a simple graph paper. This graph is known as β-absorption graph (Fig. 10.1).

As the thickness of the absorber increased, less and less β-particles are able to penetrate the metal, thus decreasing the count rate. When the thickness of the absorber (R_{max}) i.e., aluminum sheet becomes equal to or greater than the E_{max} value, no β-particle can penetrate the aluminum sheet making the count rate zero. But the graph shown in Fig. 10.1 does not seem to show zero count. In fact, the graph reveals that the count rate decreases very fast in the beginning of the measurement and then gradually it becomes almost constant. For getting the thickness equivalent to R_{max}, two tangents are drawn one before the curvature begins and the other when the count

Table 10.1 E_{max} values of some of the β-emitting radioactive isotopes and their corresponding R_{max} values

Isotope	E_{max} MeV	R_{max} mgcm^{-2}
^{3}H	0.01	80.23
^{14}C	0.155	20.0
^{32}P	1.701	810.0
^{90}Y	2.180	1065.0
^{210}Bi (RaE)	1.170	508.0

rate is almost constant (shown by dotted lines in Fig. 10.1). A perpendicular line parallel to Y-axis is drawn from the intercept of two tangents. Intersection of this line on the X-axis is taken as the maximum thickness (R_{max}) which is equivalent to the E_{max} of the β-particles.

However, a constant residual activity is always recorded instead of 0-count rate. The presence of residual activity in spite of the thick aluminum sheet is due to a phenomenon known as **Bremsstrahlung effect** and the radiation responsible for this measurement is **Bremsstrahlung radiation**. This radiation is produced when a β-particle accidentally passes through the nucleus of the absorber. β-particle then undergoes a sequence of accelerations, and eventually, it leaves the nucleus of the absorber as an electromagnetic radiation. This radiation behaves almost like an electromagnetic radiation. To stop this radiation with aluminum one would need a much thicker aluminum sheet than is required to stop most energetic β-particles. Though this effect occurs with absorbers throughout the measurements, but is noticed only when count rates are small, especially when all β-particles have been stopped by the absorber. Hence, instead of getting zero count rate, we get almost constant count rate after R_{max}. It is observed that with strong β-particles (e.g., P-32), a graph as shown in Fig. 10.1 is obtained. However, with weak β-particles, sometimes it becomes difficult to get exactly a constant value of count rate with absorbers of thickness greater than R_{max} (inset in Fig. 10.1). R_{max} calculated under this condition gives erroneous value of E_{max}. Similarly, if isotope emits γ-rays in addition to β-particles, then also deciphering R_{max} by this method becomes difficult. The magnitude of R_{max} and its equivalent value for E_{max} of some β-particles are listed in Table 10.1.

Since absorption of β-particle depends upon the thickness, as well as the atomic number of the material, thickness of metal sheet is calculated by multiplying density (mgcm^{-3}) and thickness (cm) of the metal. Thus, the unit of thickness for measuring β-absorption curve is mgcm^{-2}. Unit of thickness thus becomes independent of the type of element used to measure absorption curve. It is also obvious that for strong and weak β-particles, one may have to use thin lead and mica sheet, respectively.

When the absorption curve does not show a constant value as shown in the inset of Fig. 10.1, various other methods are used to find out the correct value of R_{max}. One way is to plot a derivative graph i.e., $d_{activity}/d_{thickness}$ versus thickness to get an inflection point in the plot. The inflection point may then be taken as an equivalent point for maximum thickness (known as R_{max}) corresponding to E_{max}. This point can be taken as the maximum range of the β-particles, provided the radioactive nuclide is a pure β-emitter. Other method is by **Feather Analysis**, which is discussed here.

10.4.2 Feather Analysis

Most of β-emitters decay along with γ-rays (*see* Decay Scheme, Chaps. 7 and 8). Due to the presence of γ-rays, it becomes difficult to decide the exact thickness corresponding to the E_{max} value. Like Bremsstrahlung radiation, γ-rays are also not absorbed by the aluminum absorber, as a result, constant count rate is observed even after the R_{max} value. Moreover, count rate is also higher than that observed with Bremsstrahlung radiation. For such cases, a Feather analysis is preferred, where the relationship between R_{max} and E_{max} is derived by comparison of β-absorption curve of the sample with that of a pure β-emitter, like ^{32}P. An absorption curve is plotted as stated earlier and approximate value of R_{max} is calculated from the method discussed earlier. From approximate R_{max} value calculated, the exact E_{max} value is obtained, with the help of empirical Feather's Eqs. (10.2) and (10.3).

$$R(\mathrm{mgcm}^{-2}) = 543 E_{max}(\mathrm{MeV}) - 160 \qquad (10.2)$$

or

$$E_{max}(\mathrm{MeV}) = \frac{R}{543} + 294 \qquad (10.3)$$

10.4.3 Graphical Absolute Method

There are a number of difficulties in determining the accurate E_{max}, by either of the methods discussed earlier, especially if the source is a weak β-emitter and associated with γ-rays. In Feather analysis, it is frequently necessary to extrapolate the line with pronounced curvature (Fig. 10.1 inset), forms of which depend strongly on the geometrical counting conditions, and the accuracy of comparing the data with the standard sample. This technique has not been discussed here, but the final equation obtained from the analysis known as Feather analysis is mentioned earlier. A graphical absolute method for determining E_{max} of β-particle developed by Barreira and Laranjeira in 1957, is preferred and discussed here.

As discussed earlier, an absorption graph is plotted between the thickness of the absorber on a semi-log graph paper, after subtracting the normal background (Fig. 10.2). A tangent, MN (Fig. 10.2) is drawn at a suitable curvature (K) of the absorption plot. Then the absorption curve is projected on to this tangent MN with horizontal lines $(AB, CD$ etc.) at the various thickness of the absorber. To every

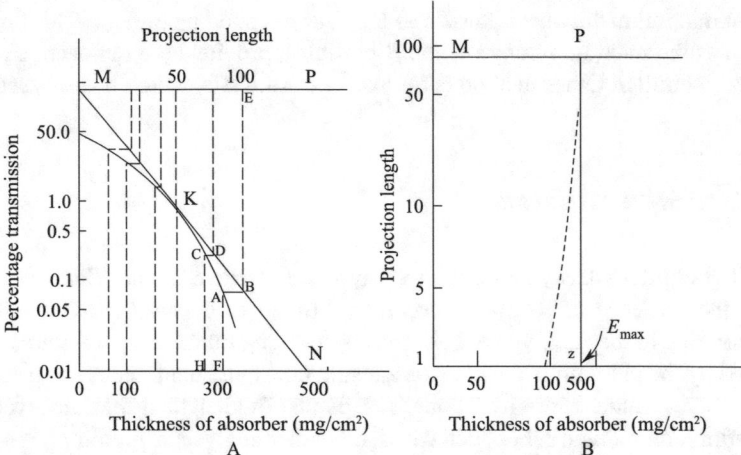

Fig. 10.2 A A typical plot of β-absorption curve. A tangent (MN) is drawn at some arbitrary point K. Line MP corresponds to projection length which is equivalent to percent transmission of β-particles through the aluminum absorber. Line AF intercepts the absorption curve at point A. AB is a horizontal line which cuts the tangent MN at point B, from where a vertical line BE is drawn. The point E is arbitrarily taken as thickness corresponding to 100% absorption. **B** A plot of projected length versus absorber thickness. At point Z corresponding to 100, a tangent PZ is drawn and the intercept on the thickness axis (Z) is taken as R_{max}

horizontal line cutting the absorption curve, we can draw vertical (AF, CH etc.) lines touching the thickness axis. Likewise, vertical parallel lines to Y-axis can be drawn such that each of these horizontal lines (AB, CD) touches a horizontal line MP. The first vertical projection line from MN to MP (i.e., BE) where the count rate is expected to be due to only Bremsstrahlung radiation, is considered as 100% absorption of β-particle by the absorber (though this may not be the real value of the R_{max}). Considering this point as 100% and the last point corresponding to zero thickness of the absorber taken as 0% (i.e., Y-axis), the entire length of this line is then arbitrarily divided into 100 equal segments (Fig. 10.2A). This MP line is called the projected length line and is equivalent to the percentage absorption of β-particles.

Now, we can generate a new set of data created on the projected line (line MP) and its corresponding thickness on the thickness axis. The absorber thickness is then plotted against the corresponding projected length MP (Fig. 10.2B). The curve thus obtained (shown by full line in Fig. 10.2B) has a smaller curvature than that obtained in the β-absorption curve (Fig. 10.1). A tangent is drawn at the point corresponding to 100% value on Y-axis to cut the thickness axis. The intercept of this line on the thickness axis (Z) is taken as the maximum thickness (R_{max}) of the β-particle for which this graph was drawn.

The advantage of this method is that we need not know exactly the position on the absorption curve where the count rate has become independent of the thickness.

10.4.4 *Determination of* E_{max}

Once the maximum absorption thickness R_{max} is determined, it is necessary to determine the corresponding E_{max} value. This can be determined with the help of a monograph (Fig. 10.3). A scale is used to connect R_{max} with a desired percentage absorption of β-particles (i.e., for external and self absorption) (Fig. 10.3). The point of intersection on the E_{max} scale shows the expected value of E_{max} for corresponding value of R_{max}. The central absorption value corresponds to the absorption of β-particles due to air, window thickness, and also the self-absorption due to the thickness of the source.

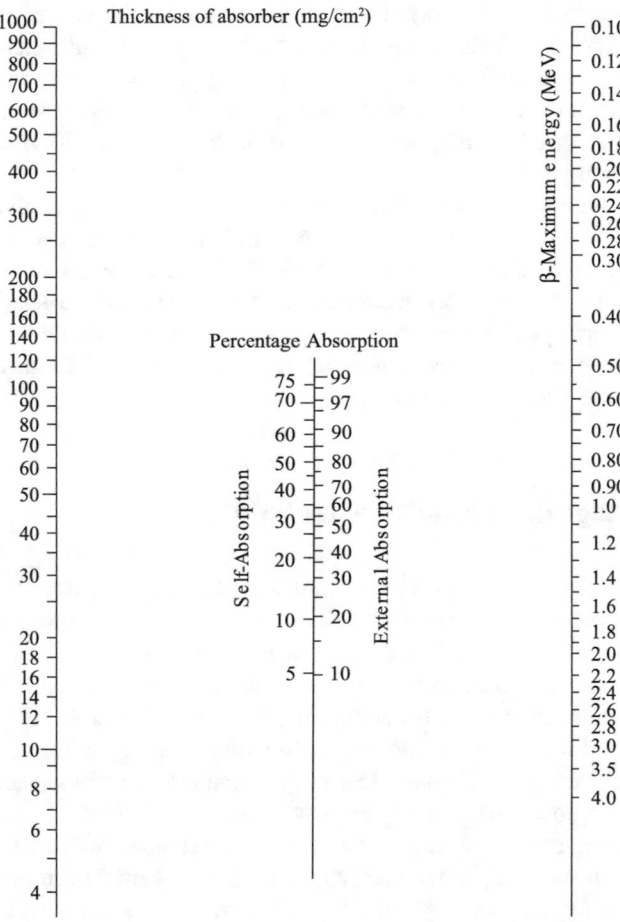

Fig. 10.3 A monograph giving the relationship between percent absorption in external absorber of known thickness and energy of β-particles (Taken from British Nucleonic Instruments, 1957)

Alternatively, the absorption curve of various β-particles of known but different energies can be taken by the method described earlier. Accordingly, their corresponding R_{max} can be calculated by either of the methods described earlier. From these results, a calibration graph between the experimentally determined R_{max} and its corresponding E_{max} energy (taken from standard table of isotopes) can be plotted, which then be used to determine the E_{max} from experimentally determined R_{max} value of a radioactive isotope.

10.5   Determination of Energy of α-Particles or γ-Rays

Energy of α-particles and γ-rays being monochromatic can be very easily determined by observing their energy spectrum with the help of the pulse-height analyzer. Measurement of activity is determined either by a gas-flow proportional counter or a scintillation counter, details of which are discussed in the previous chapters. For γ-rays, one can take help of absorption curve technique (Sect. 10.4), as well as to evaluate the energy of γ-ray.

However, there will be a need to calibrate the pulse height- analyzer in terms of energy (MeV), so that the peak position of either α-particles and γ-rays can be assigned appropriate energy in terms of MeV. This is done by taking the spectra of α-particles and γ-rays of known energy and the peak position of the spectra is plotted against their corresponding energy. These calibration graphs would be valid for the set conditions under which spectra were taken. Once the conditions are changed, the calibration curve becomes invalid.

10.6   Photographic Emulsion Technique

In a few special cases, photographic emulsion can also be a useful technique to determine the energy of α-particles from the α-tracks (i.e., distance traveled by α-particles over a photographic film/plate in lateral direction).

In this technique, a photographic plate is either impregnated with a radioactive solution and allowed to stand for some time or a solid source is kept over a photographic plate for a certain period (depending on the energy, activity, and half-life of the source) before developing. The photographic film/plate is then developed and washed. On developing, small grains with several tails (known as α-tracks) are observed. By comparing the length of a track with a standard α-track obtained with α-particle of known energy, the energy of unknown α-particle can be determined. Careful and uniform processing of the exposed film is an important part of the emulsion technique. Unless properly carried out, this technique may lead to serious errors or loss of significant results.

10.7 Half-Thickness

Since the law of exponential absorption, as stated for β-particles, is applicable to any radiation, it can also be used for γ-rays. In the equation

$$I = I_0 e^{\mu x}$$
$$\log I = \log I_0 - \mu x$$

If

$$I = \frac{1}{2} I_0$$

then

$$x_{0.5} = \frac{0.693}{\mu} \tag{10.4}$$

This suggests that the value of $x_{0.5}$ is specific for the respective energy of radiation because μ is the linear absorption coefficient (cm^{-1}) of the radiation. Thickness of absorber for which count rate becomes half of the initial value is equal to $x_{0.5}$. This thickness is called **Half-thickness**. Tables giving the energy of radiation and its corresponding half-thickness of β-particles and γ-rays are available. Therefore, once the half-thickness is determined, the energy of β-particles or γ-rays can be observed from the table, and hence, the radioactive isotope can be identified. Half-thickness, therefore, can also be used to identify radioactive isotope.

For the determination of half-thickness, we normally plot an absorption curve as described earlier, and by examining the curve, half-thickness is determined. It is worth remembering that aluminum or mica sheet absorbers are used for β-particles and lead sheet for γ-rays. This is because, γ-rays possessing more penetrating power and being energetic, would require a very thick aluminum sheet to stop. Hence, to keep the thickness to a minimum value and yet enough to stop γ-rays, absorber with higher atomic weight element like lead is used. The unit of thickness (gcm^{-2}) for this purpose is expressed by multiplying actual thickness (cm) and the density of the material (gcm^{-3}).

10.8 Half-Life Determination

Although determination of the energy of β-particles gives some information about the type of radioactive isotope present in the sample, the identity of the isotope is confirmed only after determining its half-life. Therefore, for the identification of radioactive nuclides, information regarding half-life should be obtained. The time required to decay radioactive isotope to its half activity is known as **half-life** of the radioactive isotope. The half-life of radioactive isotopes ranges from a few microseconds to millions of years. However, in the study of nuclear reactions or in radiochemical techniques, one is normally interested in short-lived nuclides, whose rate of decay could be at least measured within a span of few minutes to few days.

It is observed that decay of radioactive isotope follows the law of first order kinetic i.e.,

$$N = N_0 \exp(-\lambda \times t) \tag{10.5}$$

where

$N =$ count rate at time "t"
$N_0 =$ count rate at zero time or the time when first count was measured with respect to time "t"
$\lambda =$ decay constant.

If we take log of Eq. (10.5), we have

$$\log_e(N) = \log_e N_0 - \lambda \times t \tag{10.6}$$

If

$$N = \frac{N_0}{2}$$

which is the condition when "t" = half-life i.e., $t_{0.5}$. Then the Eq. (10.6) becomes

$$t_{0.5} = \frac{0.693}{\lambda} \tag{10.7}$$

Therefore, when ln(count rate) is plotted versus time, a linear graph is obtained with a slope of $(1/\lambda)$ and if $\log_e$ is converted to $\log_{10}$ then the slope becomes $0.693/\lambda$.

This method is applicable if the activity is measured for only one type of radioactive isotope and observations are extended over several half-lives (at least 1.5 times of the half-life of the isotope). However, when a sample contains two types of radioactive isotopes having sufficiently different half-lives, it is possible to determine the half-life of each species with reasonable accuracy. In such cases, the plot of log of activity versus time will appear as a curve (Fig. 10.4). For determining the half-life of long-lived isotope, the linear tail of the curve is extrapolated to zero time. This linear curve is taken as the activity due to long-lived isotope. Examination of this linear graph gives half-life for the long-lived isotope. The half-life of the short-lived isotope is determined by subtracting the activity of long-lived isotope from total observed count rate for the respective time. Plotting the decay curve from this subtracted data gives the correct decay curve for short-lived isotope. Examination of this linear graph gives the half-life of the short-lived isotope.

For plotting such a graph, experimentally count rate is measured with an appropriate instrument, at a number of suitable successive intervals of time. Then the logarithm of the count rate is plotted against time. A linear graph is obtained if the sample is not contaminated with another isotope of a different half-life. Half-life is found out by inspection. However, care is needed while assigning the time for which

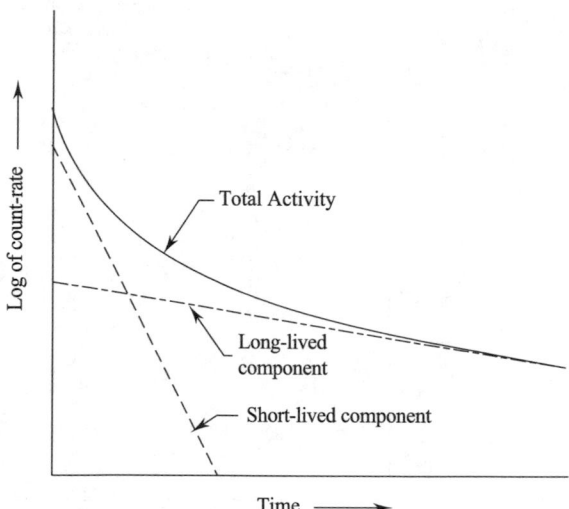

Fig. 10.4 Analysis of a decay curve obtained with a radioactive sample containing a mixture of two types of radioactive isotopes having different half-life

the count rate is to be measured. Normally, the activity is measured for a duration of 5 min (if the half-life of an isotope is more than 30 min or so). Then the count rate per unit time is calculated by dividing the observed count rate by the total time for which the activity was measured. Since the radioactive isotope would also be decaying while the measurements were made, especially for short-lived isotopes, an average time (total time for which the activity was measured divided by two) should be taken as the time for which the count rate has been recorded. While plotting the graph, this average time should be taken as the time corresponding to the count rate observed. If there is going to be no significant change in the count during the time period the activity was measured i.e., for long-lived isotopes, this averaging of time may not be necessary. In case of presence of very small amount of radiochemical impurities in the sample, though the decay plot will appear linear; but the slope may not be similar to the one we get for the pure isotope.

This method has been used very successfully to determine very short-lived isotopes (of the order of 1–5 min), provided there are facilities for counting the sample for short periods at small intervals. And if the sample is to be separated out from the mixture of isotopes, then there should also be a quick method available for separating the nuclide as well as for preparing the source for counting. For example, author measured the half-life of Thallium-208 ($t_{0.5}$) separated from thorium nitrate solution by pouring this solution slowly over a bed of freshly prepared Ammonium-12 molbydophosphate. This material absorbs specifically Thallium-208. The bed of Ammonium-12 molybodophosphate can be counted by an end-window G.M. counter. Since the half-life is 3.1 min, the G.M. counter is connected with decatron to trigger the scalar electronically to count for a set short duration after some short interval of time. For this measurement, decatron was used to trigger the scalar such that the counts were recorded for two seconds duration, at intervals of four seconds. Difficulty may arise with such type of measurements, because only a small number of

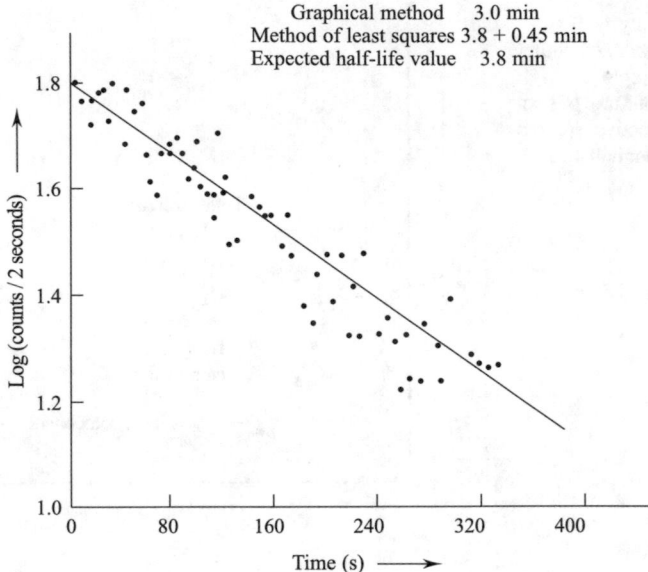

Graphical method   3.0 min
Method of least squares 3.8 + 0.45 min
Expected half-life value   3.8 min

Fig. 10.5 A plot of log of activity of Thallium-208 isotopes (separated from thorium nitrate on Ammonium-12 molybdophasphate ion exchange bed) versus time in seconds. Counting is done for two seconds at intervals of 4 s

counts would be recorded in each short interval of two seconds and the statistical uncertainty of each count would be very high. However, one could take as many readings as possible and the slope of the plot could be determined by the method of least squares. The background count in two seconds is expected to be negligible and the lost counts, due to the dead time can also be small (due to the small number of counts recorded per unit time). Hence, no corrections are needed for these two factors. It is seen from the graph that experimentally determined half-life (3.8 min) is not very much different from the expected value (3.1 min) (Fig. 10.5).

The linearity of the graph also suggests that the radioactive materials are pure and not mixed with any other radioactive impurities.

Summary

In this chapter, we have discussed methods for identifying a radioactive isotope. Two major techniques are presented: (*i*) energy determination of radiations emitted by the isotopes and (*ii*) determination of half-life of the isotopes. For determination of the energy of E_{max} of β-particles, absorption curve methods have been discussed, whereas for other radiations, like α-particles or γ-rays, normal energy spectrum or half-thickness determination has been suggested. Method to determine the half-life of the isotope is also discussed.

Chapter 11
Statistics of Counting

11.1 Introduction

Practically every book on radiochemistry deals with the statistics of numbers and it will be beyond the scope of this book to deal with this subject in much detail. However, we shall discuss the importance of statistical calculations in the detection and measurement of radioactive isotopes by taking a few examples. The measurement of radioactivity is always expressed by the number of counts per unit time. Counting is done for a longer duration and the activity is calculated by dividing the total count recorded by the time for which the counting was done. Sometimes the activity is recorded for a very short duration, especially while counting short-lived isotopes. The activities of the sample recorded by the counter are also subtracted for the background count rate to get the true count rate. Sometimes activities are measured several times to get an average count rate. It may also be required to add two or more count rates together, subtract, divide, or multiply one count rate from another count rate etc., to arrive at required results. For all these operations, a question arises as to how much count should be recorded to get the minimum error in counting the activity for a sample. Should a single count for a longer duration (e.g., 5–10 min to get a large number of count (e.g., 10,000) and divide it by time (i.e., 10,000/10) to get average count rate (i.e., 100 cpm) be taken or should one take a large number of counts of small digits (e.g., 100, 101, 99, 110, 106, 108. . .) for a shorter duration (e.g., 1–2 min) and take the mean of all counts as most accurate reading? To know which of the two methods are more accurate? if the count rate for background is low (i.e., of the order of 10–20 cpm); should the background be counted for one minute or for 1 minutes × 10 times and take an average count rate or should one count for 10 min duration to get one large count and then divide by time of counting for calculating the count rate for background activity? These questions emerge when the activity of a sample is to be reported accurately. Normally, one tends to believe that the average of the count recorded 5–10 times (with each measurement of low duration) is an accurate count. But there is a serious drawback, which can be perceived by examining the following example. A family of five persons wished to cross a river on foot. The leader of the family

© The Author(s), under exclusive license to Springer Nature Switzerland AG 2021
M. Sharon and M. Sharon, *Nuclear Chemistry*,
https://doi.org/10.1007/978-3-030-62018-9_11

before crossing the river measured the depth of water at various places (i.e., across the river) and calculated the average depth to be 90 cm. The youngest member of the family being 105 cm in height, the leader decided to cross the river with all members. To the utmost surprise to the leader, everyone except the leader was drowned, when they reached the middle of the river because the depth of the river at someplace was greater than the average depth of 90 cm. Thus, the average does not give any idea about the maxima and the minima deviations from the mean. Therefore, while representing a data, it is advisable to represent the variation in mean also.

The decay of radioactive isotope being a random process, it is extremely unlikely to obtain two consecutive count rates same value even for a long-lived isotope. Under such conditions, how do we represent an activity of a radioactive isotope accurately? In this chapter, we shall deal this question in somewhat detail.

11.2 Statistical Error in Counting

Before we can understand the significance of statistical error in counting, it may be appropriate to examine a set of counts recorded by counting a very long-lived isotope (i.e., during the course of counting, there should be no appreciable change in the activity). A radioactive Carbon-14 source was counted by a Geiger-Müller counter. Counting was done for 1 min duration for several times (Table 11.1). It is noticed that though during the period of counting of Carbon-14 (i.e., in about 45 min), there should be no change in the activity, but almost all counts have different values ranging from 9650 to 10,059 cpm. Question is which of these counts should be taken as a correct value? The average of these counts comes to 9742.5 cpm and it is accepted as the correct value.

It is customary to calculate the square root of the average value which is 98.70 cpm. We take this value as a range and then between the maximum count rate (i.e.,

Table 11.1 Count per minute recorded with C-14 sample with a Geiger-Müller counter

Counts per minute			
9858	9648	9700	9652
9744	9840	9700	9800
9659	9839	9765	9795
9645	9709	9708	9713
9800	9709	9721	9748
9679	9759	9710	9718
9809	9763	9841	9778
9840	9903	9615	9755
9740	9762	9775	9756
9753	9752	9552	9557

Table 11.2 The distribution counts as recorded in Table 11.1

Range of counts	Number of times it appears
9550–9600	2
9601–9649	3
9650–9700	3
9701–9749	11
9750–9800	13
9801–9849	5
9850–9800	1
9801–9849	1

9900 cpm) and minimum count rate 9600 cpm, we create a set of range as given in Table 11.1. We find out from Table 11.1 the number of counts appearing in each of these ranges (known as the frequency of number appearing in the range). These values are given in Table 11.2. Finally, a graph is plotted between the frequency of number appearing versus the range of count rate (Fig. 11.1). It appears from the graph that a symmetrical distribution of counts around the average count rate is obtained. This exercise reveals that the average count recorded by the G.M. counter (i.e., 78.70 cpm) though appears in counting the largest number of time, but several counts are also observed in the range of 9600–9900 cpm. Therefore, there is a need to establish a methodology so that one could represent data in a manner that reflects the variation in counts from the mean.

The distribution of counts as shown in Fig. 11.1 suggests that it is around the average count and number of time the average counts appears during the entire counting is sometimes large but certainly not all the time. Hence, if this counting is recorded once more, then unless we make a very large number of counts, the average

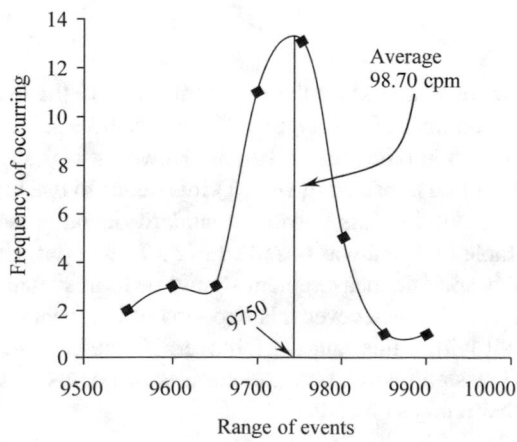

Fig. 11.1 A Gaussian type distribution of counts recorded by a G.M. counter giving the average count of 98.70 cpm

may not be a true representation of count. Then how do we represent the count such that no matter who makes the measurement, one would be able to get a similar count. For this purpose, we take the help of Gaussian distribution Law.

11.3 Gaussian Distribution Curve

When the number of events recorded in any measurement is of random nature, as is the case with radioactivity measurements, one takes the help of **Gaussian distribution law**; which suggests that if the measurements of counts follow purely statistical variation, then the distribution of counts should follow a symmetrical distribution around its mean value (Fig. 11.1). This graph is plotted from the experimental results of Table 11.2. Various useful information can be derived from this distribution curve. Before we discuss its application, it is necessary we educate ourselves with a term called standard deviation.

11.3.1 Standard Deviation

Normally, one tends to believe that the average of the count can be considered as an accurate way of representing a data. But this has a serious drawback, which can be perceived from the graph shown in Fig. 11.1 and the previous example of a family which wished to cross the river and found all members drowned except the leader of the family. Therefore, while representing a data, it is advisable to represent the variation in the mean value also, which is normally done by calculating the **standard deviation(s)**, of the mean, which is calculated from the Eq. (11.1)

$$(\sigma)^2 = \frac{\sum\limits_{i}^{n}(x - x_i)}{(n - 1)} \tag{11.1}$$

where x_i and x are the ith count rate and the average count rate, respectively. n is the number of observation. For an accurate standard deviation calculation, the value of n should be greater than 30. However, if the value of n is less than 30 then in the Eq. (11.1), instead of $(n - 1)$ it is better to use only n.

Applying this equation standard deviation was calculated for count shown in Table 11.1 and was observed to be 77.92 cpm. Hence, the count rate as observed in this specific measurement should be represented as 9570 ± 77.92 cpm. This would mean that whenever this experiment is repeated by any other person, his data will fall within this range. Moreover, if results as mentioned in Table 11.1 follows the Gaussian distribution, the number of counts as shown in Table 11.1 should follow the trend as given here

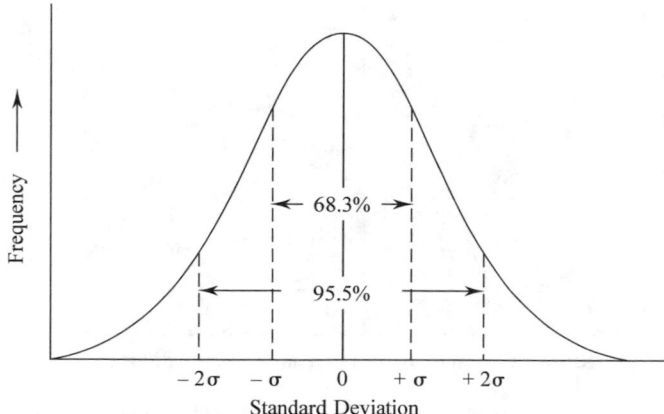

Fig. 11.2 An ideal Gaussian distribution curve is shown here. Mean value is taken as the zero and variation of counts with respect to mean values is shown. 68.3% and 95.5% are expected range of the distribution if there is no other error involved in the counting system and the radioactive material does not decay during the course of counting

- mean $\pm 1\sigma$ 68.3% of total count should fall in this range
- mean $\pm 2\sigma$ 95.5% of total count should in this range
- mean $\pm 3\sigma$ 99.9% of total count should fall in this range.

However, it will be noticed that the data shown in Table 11.1 and the graph plotted from these data do not fall in line with the expected Gaussian distribution curve (Fig. 11.2). As a result of this, even the expected value of mean $\pm 1\sigma$ (i.e., 9570 $\pm$ 77.92 cpm) contains more than 68.3% of the total data. There could be various reasons for getting such deviation from the expected result which is discussed in the next section.

The advantage of this Gaussian distribution analysis is that by plotting a Gaussian distribution of radioactivity of a long-lived isotope, one can confirm whether the counting system is operating properly. If the counting system is operating perfectly and recorded counts are following the Gaussian distribution curve, then a symmetrical distribution curve is obtained as shown in Fig. 11.3C. But if the curve takes the form as shown in Fig. 11.3A or B, then one should examine the counting system or even the counter may be behaving erratically.

There are, however, some practical difficulties in either plotting this curve or calculating the standard deviation accurately, especially when the radioactive isotope is either a short-lived one or the activity is very low. In such cases, a single count is recorded for a duration in which the decrease in the activity due to decay the process is negligible and the square root of this value is taken as an approximate value of the standard deviation.

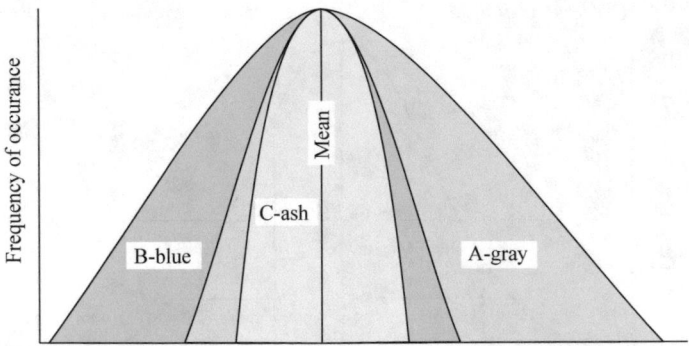

Range (deviation from mean in term of standard deviation)

Fig. 11.3 Gaussian curves **A** showing unsymmetrical curve with a large number of data having a value greater than mean, **B** showing unsymmetrical curve with most of the number being smaller than mean and **C** showing a symmetrical Gaussian curve with respect to the mean

11.3.2   Advantages of Standard Deviation Calculation

The standard deviation is like a representation of an error involved in the recording of a count. Hence, one can calculate the percent error associated with measuring various numbers (Table 11.3). Thus, from the data shown in Table 11.3 it can be concluded that if the total number recorded in one measurement is 10, then due to statistical nature of the number, the error is as high as 33.3% and when the total number is 10,000 then the error associated is only 1%. It should also be remembered that this error can be improved only by counting larger number of counts. Hence, the average of a small number does not mean that the number recorded is more accurate than one single large number. Even when an average value of 1000 recorded for 100 times, each of number recorded will contain an error of 3.3%, hence the average of 1000 recorded number will also contain 3.3% error. On the other hand, a single record of 10,000 number will have an error of only 1%. In other words, lower the number of counts observed, higher is the error associated with it. Hence, in radiochemical

Table 11.3 The percent error in counting various numbers is shown here

Number of counts recorded in one single observation	Expected deviation from mean i.e., square root of mean	% error in counting such numbers
10	3.3	33.3
100	10.0	10.0
1000	33.3	3.3
10,000	100.0	1.0
100,000	333.3	0.3
1000,000	1000.0	0.1

work, one should try to measure the activity for the longest possible period so that the total counts recorded is as high as possible. Certainly, we shall face problems in recording the large number of counts with either low active materials of long-lived isotopes or short-lived isotope. Because short-lived isotopes with low activity can not be counted for a longer duration as it will decay while counting. In such cases, one has to make a compromise. Similarly, when counting a source of low activity of long-lived isotopes, one may need several 100 h to counts even 1000 counts. Under these conditions also one has to make a compromise.

11.3.3 Representation of Activity

The measurement of radioactivity is always expressed by the number of counts per unit time. Counting is done for a longer duration and the count rate is calculated by dividing the total count recorded by the time for which counting was done. The activities of the sample recorded by the counter are also subtracted for the background count rate to get the true count rate. One may also need to add two or more count rates together or subtract or divide or multiply one count rate from another count rate etc., to arrive at required results.

The next question that arises is when counts are to be multiplied, or divided, added etc., then how can we represent the standard deviation of the final answer? In the next few sections, we shall devote time to calculate the standard deviation for these different conditions.

11.4 Sums and Differences of Counts

In radiochemical works, the recorded activities of two samples are either to be subtracted or added to get the final results. The subtraction of background from the sample is a very common example. In such cases, the standard deviation of final results is calculated by the Eq. (11.2). It is assumed that the count to be subtracted is less than the count from which it is to be subtracted, which is also the case with the background count. Because the background count is normally very low compared to the actual count rate of the sample. We can define

$$u = x + y \quad \text{or} \quad u = x - y$$

where x and y are two types of counts (i.e., observed count being x, the background count being y and net count to be calculated is u). It will be seen that the standard deviation of the final result (σ_u) is the same, whether the counts are subtracted or added and is given by Eq. (11.2)

$$(\sigma_u)^2 = (\sigma_x)^2 + (\sigma_y)^2 \tag{11.2}$$

where σ_x, σ_y and σ_u are the standard deviations for samples x, y and u respectively. We can take one example to explain this calculation.

If 2051 cpm and 621 cpm were recorded for two samples, then what would be the standard deviation for the final result?

The actual count recorded is $x = 2051$ cpm and the other count recorded (which could be background count) is $y = 621$ cpm. Hence, the sample's count $u = (x - y) = 1430$ cpm. If we assume that these counts are approximately equal to mean (because normally we do not carry out the measurements several times to calculate the mean count), we can calculate the standard deviation of the final result as follows:

$$(\sigma_u)^2 = (\sigma_x)^2 + (\sigma_y)^2 = 51.69$$

The final result should, therefore, be represented as $u = (x - y) \pm \sigma_y = 1430 \pm 51.69$ cpm. This calculation, thus, suggests that if this sample is counted many times, and if there is no error in counting other than the statistical error, then 68.3% of entire recorded counts should lie within the limit of 1378.31 (e.g., $1430 - 51.69$) to 1481.69 (e.g., $1430 + 51.69$).

11.5 Multiplication or Division to Recorded Count

Multiplication or division could be either with two sets of counts or by a constant factor. The calculations for both are done differently. These are discussed in two sections separately.

11.5.1 Division by a Constant Factor

Many times the counts are to be multiplied by a constant factor or divided by a constant factor. The standard deviations for the final result for such cases are expressed by Eqs. (11.3) or (11.4).

If we define $u = Ax$, where A is a constant factor to be multiplied by the observed count x, then the standard deviation of the final result is given by Eq. (11.3)

$$\sigma_u = A\sigma_x \tag{11.3}$$

In other case, where $u = x/A$, i.e., the count x is divided by a constant factor A, then the standard deviation of the final result is given by Eq. (11.4).

$$\sigma_u = \frac{\sigma_x}{A} \tag{11.4}$$

In both the Eqs. (11.3) and (11.4), it is assumed that the multiplication or division is done by a constant factor which does not alter the relative error. We can take one example to illustrate this calculation.

2015 counts were recorded in 15 s. What would be its standard deviation when counts are expressed in cps and cpm?

Counts recorded in 15 s = 2015

$$\text{count rate } u = 2015/15 \text{ counts/s} = 134.33 \text{ cps}$$

The standard deviation (σ_u) is

$$\sigma_u = \frac{\sigma_x}{t} = \frac{\sqrt{2015}}{15} = 2.99\text{s}^{-1}$$

When the count is to be expressed in cpm

$$u = \frac{60 \times 2015}{15} = 8249 \text{ cpm}$$

Then

$$\sigma_x = \sqrt{2015} = 44.89$$

The standard deviation (σ_u) of the final result becomes

$$\sigma_x = \frac{\sigma_x \times 60}{15} = 179.55$$

Count/minute should be expressed as 8239 ± 179.55 cpm. Statistical calculation thus gives a yardstick to find out whether the counting procedure and counting system is operating within the limit of statistical variation.

11.5.2 Multiplication by Another Count

During some calculations recorded activity may have to be multiplied or divided by another recorded counts. For such cases, the standard deviation of final result (σ_u) can be calculated as follows:

1. **For a case when $u = xy$**

   The standard deviation (σ_u) is given by the Eq. (11.5)

$$\left(\frac{\sigma_u}{u}\right)^2 = \left(\frac{\sigma_x}{x}\right)^2 + \left(\frac{\sigma_y}{y}\right)^2 \tag{11.5}$$

2. **For a case when** $u = x/y$

The standard deviation (σ_u) is given by the Eq. (11.6)

$$\left(\frac{\sigma_u}{u}\right)^2 = \left(\frac{\sigma_x}{x}\right)^2 + \left(\frac{\sigma_y}{y}\right)^2 \tag{11.6}$$

Thus, in either of these two cases, the standard deviation (u) is calculated in the same fashion. We can take an example to explain these two calculations.

If the activity of two sources is 15826 and 6721 cpm, what would be the standard deviation of the ratio of these two activities?

Counts for the source 1, i.e., $x = 15826$ cpm. Hence, its approximate standard deviation

$$\sigma_x = \sqrt{15826} = 125.80$$

Likewise, counts for the source 2 (i.e., y) being 6721 cpm, its approximate standard deviation

$$\sigma_y = \sqrt{6721} = 81.98$$

Ratio activity

$$u = \frac{15826}{6721} = 2.35$$

The standard deviation as per the Eq. (11.6) is

$$\left(\frac{\sigma_u}{u}\right)^2 = \left(\frac{125.8}{15826}\right)^2 + \left(\frac{81.98}{6721}\right)^2$$
$$= 0.02015$$
$$\left(\frac{\sigma_u}{u}\right) = 0.14$$

Thus,

$$\sigma_u = 0.14 \times 2.35$$
$$= 0.33$$

The final result should be expressed as 2.35 ± 0.33 cpm.

Summary

Radioactive decay being a random process, it is important to represent results along with the standard deviation. Though it is desirable to count large number of digits within one observation to get an accurate value of the standard deviation, one may have to make a compromise while dealing with short-lived isotopes with low activity, as well as with long-lived isotopes of low activity. In both cases, square root of single observation may be taken as approximately equal to the standard deviation. It is also obvious from these discussions that mere representation of mean is not the best way to represent the data. When digits recorded are to be multiplied, divided subtracted or added to another either fixed or variable number, the standard deviation of the final result can also be expressed by applying the appropriate equations as discussed in this chapter.

Chapter 12
Health Hazards and Protection

12.1 Introduction

While handling radioactive isotopes, mere knowledge of radiochemical techniques to perform some specific chemical reaction/separation, or to detect, measure the activity etc., are not enough. It is equally important to be aware of the health hazards associated with radioactive isotopes. Moreover, it is also important to realize that the laboratory where such work is expected to be carried out, should take care of the disposal and contamination problems associated with some of the hazardous radioactive isotopes like ^{137}Cs. Unlike a chemical laboratory, the radiochemical laboratory has to differentiate between the radioactive zone and non-radioactive zone and many cares related to the safety of the personnel working in the radiochemical laboratory are to be taken. Before we can appreciate the problems in dealing with the radioactive materials, it is necessary to understand the reasons, which cause the health hazard. In this chapter, therefore, we shall be mainly concerned with these aspects and discussions would be geared to explain the various factors necessary to be considered before launching to work with radioactive isotopes.

12.2 Biological Effect of Radiation

Why are radioactive isotopes dangerous to human beings? Before we can appreciate the problems of hazards, it may be useful to understand something about our human body and the chemical characteristics of human cells.

We have discussed in the earlier chapters that any type of radiation e.g., particulate radiations, neutrons or electromagnetic radiation produces ionization and chemical changes in their passage through the matter. In a biological system, this ionization can cause damage, directly by disruption of chemical bonds in the cell, both in the **cytoplasm** and the **nucleus**. A cell of a human being is broadly described as having a nucleus around which another organic material known as cytoplasm is present.

Damage to the cytoplasm is up to a point recoverable, provided the rate of damage is less than the rate of recovery, whereas the damage to the nucleus could cause mutation of genes in the chromosomes, which exist in a specific number of pairs i.e., 23 in the human body. Mutation may change the characteristic of the cell or may cause cell death. Sometimes the order of the genes is also changed, due to breakage in the chromosomes. These may affect human characteristics e.g., colour of eye, fall of hair etc. In short, radiation of any kind or energy causes a permanent damage when it interacts with the nucleus.

Some tissues are more radio-sensitive than the others, such as bone marrow, in which the red and the white corpuscles are produced. There are also tissues which are little affected by radiations, such as skin, nail. Hence, hands, feet, ankles or other parts of the body whose main constituents are skin and bone will be less affected by radiations. In a tracer laboratory, especially in experiments in which nuclear radiation with either neutron generator or laboratory neutron sources are used, the danger to the body may be due to either external radiation or internal radiation (i.e., due to ingestion of radioactive substances). The extent of the hazard depends upon the type of radiation and the type of exposure.

While talking about the hazards of radiation to the human body, we shall have to address it in two ways: (i) Are the radiations causing the damage to the cell externally such that the rate of physiological equilibrium is affected. (ii) Or are these radioactive isotopes chemically similar to chemicals, which follow the metabolic path and thus retained by the human body either at a specific organ or throughout the body causing damage to the physiological growth/decay process? The former effect is related to the nature and energy of the radiations emitted by the radioactive isotopes, while the latter is associated with the chemical nature of the radioactive isotopes. In a biological system, therefore, the ionization of organic components of the cell by the radiations can cause damage to the chemical composition of the cell, directly by the disruption of chemical bonds in the cell, both in the cytoplasm and the nucleus. Moreover, radiation of any kind or energy when interacts with the nucleus of the human cell can cause permanent damage to the cell.

The effect of radiations on the human body can, therefore, be of two types: (i) effect of radiation due to external radiations and (ii) effect of radiation of the isotopes which are present inside the body due to some ingestion. The damage due to external radiation depends naturally on the type of radiation and its energy. The internal damage will depend upon the placement of the radioactive isotopes in the human body. For example, some tissues of the human body are more radio-sensitive than others, such as bone marrow, in which red/white corpuscles are produced and if the isotopes are present in the vicinity of theses tissue, the damage would be of great significance.

In a radiochemical laboratory, we, therefore, are likely to receive externally, radiations from a source like neutron generator or laboratory neutron sources or radiations due to the handling of radioactive samples. These radiations either may be capable of penetrating our body and causing damage to our cells within the body or may be stopped by the skin of our body causing a local external damage to the body. The latter is possible with sources like weak β-emitters or α-emitters. The damage done to the body by weak β-emitters or α-emitters are to the somatic tissues. However, sometimes radioactive isotopes may be taken inside the body through some cut in the body etc., then the extent of damage will depend upon the nature of the isotopes. The effect of these damages may be carried over to next generation, and are known as genetic damage. Hence, the damages caused by the radiation may be classified as (i) external exposure and (ii) internal exposure. We shall now discuss these two aspects, in the next section.

12.3 External Exposure

The extent of external hazard depends on the type of radiation, because of their difference in specific ionizing power and penetrating power. We have seen in the previous chapters that X-rays and γ-rays have very high penetrating power, hence can penetrate through the skin into the interior parts of the body or can come out of the body without being absorbed. These properties have some advantages and some disadvantages. If these radiations are not stopped by the human body, the chances of causing ionization to the human cells are low, because most of them will come out of the body before they have the chance to transfer their energy to the human cell. However, if these radiations are stopped by hard tissue of the body like bones, then the damage can be of large magnitude.

In the medical profession, this specific property of radiation has proved a very useful tool to reveal much information about the human body like condition of bones, obstructions created in some of veins etc. For this purpose, a specific part of the human body is exposed with soft X-rays of low intensity by keeping an X-ray film behind the body. The X-ray film after developing shows a white spot where X-rays were obstructed by the human body. The examination of the X-ray film thus can give lots of information, which are useful for medical diagnosis.

On the other hand, particulate radiations depending upon their energy (like energetic β-particles, being less penetrating than X-rays or γ-rays) can produce a greater amount of external damage to human tissue, because these radiations can be stopped by the skin or the mussels of human body. Weak β-particles or α-particles, can be stopped by the skin and hence they can do almost no external damage to the body except for the skin of the human body.

Neutrons are very hazardous because of their high penetrating power, and monitoring of neutrons is also difficult. In fact, neutrons are considered to be the most hazardous of all types of radiations.

In a radiochemical laboratory, however, we are not likely to use large quantity of γ-emitting isotopes and hence chances of encountering external damages due to these radiations are very unlikely. However, if we are using the neutron sources or γ-sources for carrying out some experiments, then one has to be extremely careful while working in the area where these sources are kept. We must monitor regularly the amount of radiations received by the body so that we do not exceed the permissible level of dose.

In conclusion, it can be said that, in spite of having high penetrating power, X- or γ-rays can still do some damage to the human body. Among the particulate radiations, high energy β-particles can do localized damage to the human body. Therefore, we shall have to prescribe a minimum permissible amount of electromagnetic radiation, which can be tolerated by the human body.

12.4 Internal Absorption

Internal damage to the human body is caused by the radiation emitted from radioactive material taken into the body by inhalation, ingestion, absorption through the skin or through an open wound etc. γ-emitting isotopes if taken within the body, produces relatively less damage to the cell, because most of the radiation comes out of the body (due to its high penetrating power), without causing any appreciable damage to the tissue present inside the body. On the other hand, α-particles or β-particles, having less penetrating power can do large amount of damage to the cell present inside the human body. α-particles, in fact, the most hazardous, because they produce a high degree of localized damage. Hence, as a rule, maximum precautions should be taken to avoid ingestion of any type of radioactive materials inside the human body. However, the physiology of the human body is such that, most materials (either radioactive or non-radioactive) taken by the human body, are also discarded or excreted out of the body by the ongoing biological process. The body retains some materials for a longer duration and some for a short duration. This is measured by the **biological half-life**, which is the **time needed by the biological process to eliminate the material to its half concentration**. This biological process, thus, helps us to eliminate the radioactive isotopes from the body.

Nevertheless, we have to be careful working with radioactive material emitting especially weak β-particles or α-particles and its chemical nature. There are certain radioactive isotopes which are concentrated in a particular part of the body and cause a localized damage. For example, ^{90}Sr and ^{89}Sr being chemically similar to calcium, are carried along with calcium into the body and deposited in the bone. Because calcium is used for the bone formation, and thus ^{90}Sr and ^{89}Sr are retained within the body for a very long period in the bone. Since bone marrows are responsible for the production of red/white blood corpuscles, any radioactive material following the path of bone fixation process will create extensive damage to the activity of bone marrow cells.

There are also radioactive isotopes which are diluted throughout the body and cause a **genetic hazard (a hazard caused to the human cell such that its effect is carried over to the next generation)**. For example, ^{137}Cs, which follows the chemistry of potassium, is taken over the entire body causing a genetic hazard. Genetic hazard is the most dangerous type, because its effect is observed in the offspring and hence can be carried over to the next generation, as is being observed with people who were affected in Japan, due to the explosion of the atomic bomb in Hiroshima and Nagasaki. Therefore, in principle, the internal hazard due to radioactive isotopes is highly dangerous. Since we have to work with such radioactive materials, based upon experience and chemical nature of the isotopes, they have been classified into four categories:

Category	Class	Examples
Very highly toxic	class-I	^{90}Sr, ^{242}Cm etc.
Highly toxic	class-II	^{45}Ca, ^{89}Sr, ^{140}Ba etc.
Moderately toxic	class-III	^{82}Br, ^{137}Cs, ^{137}Ba etc.
Slightly toxic	class-IV	^{7}Be, ^{51}Cr etc.

Therefore, unlike the external hazard, the damage caused by the radioactive isotopes present inside the body depends on the chemical nature of the isotopes, type of radiation emitted by the isotope, the energy of the radiation, biological half-life and its radioactive half-life. Therefore, half-life of radioactive isotopes for the biological process, is expressed in terms of **effective half-life** which takes into account of these two types of half-lives. The effective half-life is calculated from Eq. (12.1)

$$\frac{1}{t_{\text{eff}}} = \frac{1}{t_{\text{rad}}} + \frac{1}{t_{\text{biol}}} \tag{12.1}$$

where, t_{eff}, t_{rad} and t_{biol} are the effective half-life, half-life of the radioactive isotope, and the biological half-life of the radioactive isotope, respectively.

Considering these factors and the experiences, which we have gained over the period in handling radioactive isotopes, International Atomic Energy Commission has prescribed the maximum limit, which a personnel working with radioactive isotopes can handle for these four classes of materials. These are listed here:

Category	Permissible level of activity
class-I	1 mCi
class-II	10 mCi
class-III	100 mCi
class-IV	1000 mCi

12.5 Units of Radiation Dose

In radiochemical experiments, we are normally interested in the number of particles emitted per unit time by the radioactive isotopes. Likewise, in setting a limit of the radioactive materials to be handled safely by personnel, we again have to decide the limit based upon the number of particles being emitted by the isotopes. Thus, in either of these two cases, we need to know the number of particles being emitted by the isotopes per unit time. Therefore, the unit to measure the radiation dose must be related to the number of particles emitted by the isotopes per unit time. Moreover, for defining the limit for external hazard though, we need to measure the number of radiations emitted per unit time, but for internal hazards, we need to define the unit of radiation, which is related to the amount of material and the number of radiation emitted by the isotopes. Considering these two aspects of the requirements, the unit of radiations are classified broadly into two categories (i) unit of rad (or rem)/unit time and (ii) curie.

12.5.1   Curie

When we wish to know the amount of radioactive material present in the sample so that we can get some idea about the number of radiations, which could be emitted by the isotope per unit time, the unit for such purpose is **Curie**, the name given in the honor of Curie family who discovered the application of radioactivity.

This unit was established by following the decay of Radium-226 isotope. The atomic weight of radium is 226.03. Therefore, if pure Radium-226 is separated then at zero time of its formation (i.e., No), 1 gram of Radium-226 will contain $(6.022 \times 10^{23}/226.03) = 2.664 \times 10^{21}$ atoms. The half-life of Radium-226 being 1600 yr, hence its decay constant (λ) is equal to $[0.69315/(1600 \times 3.156 \times 10^7)] = 0.1373 \times 10^{-10}$ per second, 3.156×10^7 being the number of seconds in a year. Thus, disintegration rate of 1 gram of Radium-226 is given by $dN_0/dt = 0.1373 \times 10^{-10} \times 2.664 \times 10^{21} = 3.66 \times 10^{10}$ α-particles (or radon atoms) per seconds, α-particle being the type of radiation emitted by ^{226}Ra and radon being the decay product of this isotope. Thus curie (Ci) is defined as the amount of radon (the decay product of Radium-226) in equilibrium with 1 gram of Radium-226. Based on this calculation, a general definition of one curie (1 Ci) is the activity of a radioactive substance which decays (i.e., emits radiations) at a rate of 3.66×10^{10} disintegration per second. Curie is thus the quantity of radioactive material which emits 3.66×10^{10} disintegration per second. The advantage of this unit is that we can use this quantity to establish the efficiency of counting process, or the efficiency of chemical separation techniques etc.

12.5.2   Rad

When our interest is in knowing the amount of radiations received externally by the person per unit time, the unit of this amount is referred to as **rad**. It is defined as the **dosage of radiation that will impart 100 ergs of energy to each gram of matter through which the radiation passes**. The rad is being replaced by its SI equivalent, the gray (Gy) defined as 1 J/kg. The two units are thus related by 1 Gy $=$ 100 rad. These units are used to specify the maximum permissible level a person can receive, which may not cause alarming damage to the human body.

12.5.3 Rem

Another unit of radiation is **rem**, which is used for biological system. The need for a separate unit for the biological system arises because the amount of damage made to the tissue of the biological system depends upon the extent of absorption of the radiation during its path in the biological system. For example, α-particles of same energy as that of γ-radiation will do much extensive damage to the biological tissues. We have seen earlier that when radiations pass through the biological tissues, they create chemical alteration to the biological molecules. The severity and permanence of these changes are directly related to the local rate of energy deposition along the particle track, known as **linear energy transfer** L. Obviously, α-particles will have a larger value for "L" as compared to γ-rays, as the latter cannot transfer energy to the biological molecule as effectively as the former.

Thus, the magnitude of damage to the biological tissues would depend upon the value of "L". Hence, for biological molecules, we have to consider two aspects of the radiation, its magnitude of radiation being received by the tissue (i.e., "D" in terms of rad/h, as defined earlier) and a quality factor (i.e., "Q" a dimensionless factor) which is related to the magnitude of "L" and is related to the nature of radiation being considered. The product of these two quantities i.e., "D" and "Q" can, therefore, be taken as the real measure to express the unit of radiation for the biological system. The product of these two items is defined in terms of **rem**. In other words, for a biological system, the effective quantity of radiation which will damage the system has to be multiplied by a factor arising of the fact that radiations with high specific ionization properties (like α-particles) will cause damage to the large number of biological tissues of the body as compared to radiation of less specific ionization properties (like γ-rays), though both radiations may have the same energy.

12.5.4 Maximum Permissible Level

Our body is exposed not only to radiations due to radioactive isotopes present in the laboratory, but is also continuously exposed to small intensities of ionization radiations from cosmic radiation or from radioactive substances present on the earth and in the air. The scientific and medical investigations of these radiations on the body and the effect of the higher dose rate on the animals has shown that there is a tolerance level of radiation dose, due to either external exposure or internal ingestion. The former level is called **maximum permissible level of external radiation**, and is defined as the dose rate or level of radiation below which no permanent physiological changes are thought to be likely. The level is not a fixed value and is liable to be changed, by the International Commission for Radiological Protection (I.C.R.P.), should any evidence become available which indicates such alteration.

12.5.5   Maximum Permissible Level of External Radiation

According to I.C.R.P., the total exposure dose accumulated over several years of any age over 18 shall not exceed that given by the relation

$$D = 5(\text{Age} - 18) \text{ rem} \qquad (12.2)$$

where D is the tissue dose in unit of **rem**. This relationship implies that on an average, a dose rate of 5 rem per year or 0.1 rem per week is a permissible dose. Over a shorter period, this latter dose-rate may be exceeded, provided that it is not greater than 3 rem during any period of 13 consecutive weeks. The long term average dose in hands, forearms, feet, and ankles should not exceed more than 1.5 rem/week, and in addition, the maximum accumulated doses of 20 rem in 13 consecutive weeks. In the tracer laboratory, the working dose should not exceed 2.5 rem/hr.

12.5.6   Maximum Permissible Body Burden

The maximum permissible level for internal ingested radioactive isotopes can also be set on the experimental evidence obtained so far. This level is called the **maximum permissible body burden** and it is customary to define this with reference to the organs of the body. This level depends upon the radioactive isotopes concerned i.e., on its toxicity and the effective half-life of the radioactive isotope.

The **maximum permissible body burden** is defined as that **level or amount of the isotope, which if ingested into the body, will produce no appreciable health hazard**. This level, unlike the maximum permissible level of external radiation, is liable to change. The unit of maximum permissible body burden is the curie, because here one is interested in the amount of radioactive isotopes taken in by the body. A few important values are listed in the table (taken from Health Physics, *Official Journal of Health Physics Soc.*, Vol. 3, June 1960)

Nuclide	Organ of reference	Maximum body burden
Iodine-131	Thyroid	0.77 mCi
Cesium-137	Total body	50 mCi
Strontium-90	Total body	30 mCi
Bromine-82	Total body	10 mCi

Since it is extremely difficult to measure the amount of radioactive isotopes ingested in the body, it is always essential to avoid any possible way of ingestion, such as through the wound or cut in the hands, skin etc., while handling the radioactive materials. These levels apply, not only to people occupationally exposed to radiation, but also to the general population.

12.5.7 Dose Rate Calculation

The dose-rate can be calculated by using a simplified mathematical equation, but these calculations give approximate results. In calculating the dose-rate, the decay scheme of the radioactive isotope must be known. This helps in finding out the different types of radiations emitted, their energy, and their % abundance. With isotopes giving several radiations with different % abundances, the total energy is calculated which is the sum of products of energy and the % abundance for each type of radiation separately and this value should be considered in the calculation of dose rate. If more than one radiation is emitted per disintegration, the dose-rate is multiplied by the number of radiations emitted.

12.5.7.1 γ-Rays or X-Rays

For γ-rays or X-rays the number of rad/hr obtained at a distance of "d" feet from an unshielded point source of γ-rays within the range of 0.2–1.5 MeV may be calculated from the source activity using the equation

$$\text{Dose-rate} = \frac{N}{4\pi d^2} \times T_d \times E_{\text{abs}} \times 5.766 \times 10^{-5} \text{ rad/hr} \qquad (12.3)$$

where

N = number of radiations per disintegration.
d = distance between the source and point of interest (in feet).
T_d = transmission factor (it is usually unity for γ-rays and less than one for β-particles).
E_{abs} = energy absorbed per cm path in tissue. This value may be obtained from Nuclear Handbook.

If the transmission factor and the energy absorbed per cm of path in tissue be known, then this equation can be used either for γ-rays or β-particles. However, an approximate equation usually used for the calculation of dose-rate is

$$\text{Dose-rate} = \frac{6CEN}{d^2} \text{ rad /hr} \qquad (12.4)$$

where

C = source strength in curies,
E = energy of γ-rays,
N = number of radiation per disintegration, and
d = distance from the source in feet.

Likewise, a very simple equation is used for the calculation of the dose-rate for β-particles:

$$\text{Dose-rate} = 300\ C\ \text{rad/hr at one foot}$$

where, C is source strength in curies.

These calculations of dose-rate can be explained by considering an example of a radioactive isotope Bromine-82, (*see* the decay scheme of ^{82}Br) which gives several γ-rays and one β-particle of energy 0.46 MeV. From the decay scheme of Bromine-82, it will be noticed that each γ-rays has its own % abundance and the total energy of γ-rays considering their abundances comes to about 2.697 MeV. Though not a single γ-rays of Bromine-82 has such high energy, the body will experience the same effect as if one γ-rays of energy 2.697 MeV were causing the damage. Hence, in the calculation of dose-rate, 2.697 MeV should be considered. Consequently, if the source activity is 1.7 mCi (millicuries), the dose-rate due to γ-rays at one foot would be

$$D = 6CEN$$
$$= 6 \times 1.7 \times 10^{-3} \times 2.697 \times 1$$
$$= 27.5 \times 10^{-3}\ \text{rad/hr.}$$

Calculation for dose-rate for β-particles would require only the source strength i.e., 1.7 mCi. Hence,

$$\text{Dose-rate} = 300C\ \text{rad/h at one foot}$$
$$= 300 \times 1.7 \times 10^{-3}$$
$$= 510 \times 10^{-3}\ \text{rad/h}$$

Thus total dose-rate received by the person working with ^{82}Br isotope of 1.7 mCi at one foot distance is sum of these two dose rates i.e., 510 mrad/hr + 27.5 mrad/hr = 537.5 mrad/hr.

It is noteworthy, that although the dose-rate due to β-particles in the calculation seems to be the highest, it has very much less effect on the body. Since most of the β-particles (E_{max} = 0.46 MeV) will be absorbed by the glass container, the lead-pot etc., very few fractions of β-particles would be able to reach the body at one foot distance. On the other hand, the Bremsstrahlung radiation may perhaps be appreciable. Nevertheless, γ-rays will be received by the body without any considerable loss, unless lead shielding has been used. In other words, in practice the dose-rate at one foot will not be 537.5 mrad/hr, but definitely more than 27.5 mrad/hr. This calculation also points out the difficulties in calculating theoretically the dose rate. Therefore, it is always better to experimentally measure the dose rate to get the real dose-rate. Nevertheless, theoretical calculation does give some idea about the magnitude of dose likely to be received by the isotope.

12.6 Protection from External Hazard

While working in radiochemical laboratory, it often happens that the dose-rate of the sample is higher than the maximum permissible level, especially while dispensing the highly active materials (like ^{82}Br). In such cases, the dose-rate is cut down by one or more methods discussed in foregoing sections.

12.6.1 Distance

The intensity of radiations reaching the object is inversely proportional to the square of the distance between the source and the object. Therefore, distance acts as a good shielding. In the practical work, it is not possible to handle the radioactive material at a long distance unless there are facilities, like remote control systems. However, tweezers of six inches long are used to handle the radioactive sources in order to reduce the dose-rate to some extent.

12.6.2 Shielding

It is possible to reduce the hazards of nuclear radiation by using a shield. Usually, lead bricks are used for shielding γ-rays. One can use relatively thinner lead bricks as compared to aluminum bricks to stop the β-particles to the same magnitude because the latter element has low atomic weight.

Since γ-radiation may be scattered in all directions, care must be taken to see that there is an adequate shielding from all the directions to prevent scattered radiations reaching the person handling the radioactive material. However, for β-particles, elements of low atomic number such as aluminum, glass, polythene sheets etc., are used for shielding purposes, because though the high atomic number nuclei would stop the β-particles more effectively with a thinner sheet, but these materials increase the Bremsstrahlung radiation for which we may need thicker lead bricks. Hence, it is better to avoid the generation of Bremsstrahlung radiation by avoiding high atomic weight elements. Unfortunately, there is no simple satisfactory method for calculating the shielding thickness, taking due account of the scattered radiation. The usual method is to check the dose-rate, at various distances from the shielding, by a hand monitor, and if the dose-rate is higher than maximum permissible level, more shielding has to be used.

12.6.3   Time

Time is an alternative method to minimize the dose-rate. A dose higher than the maximum permissible level may be tolerated for a shorter duration, provided the average dose for the period (e.g., one week), is below the maximum permissible level. However, this should not become a regular practice but may be used when other methods of shielding are unsuitable.

12.7   Instruments for Detection and Measurement of Radiation

In a radiochemical laboratory, it is necessary to keep a continuous watch on the level of activity due to radiations, so that if there is any sudden change in activity level in the laboratory due to any accident or spillage etc., necessary precautions could be taken, immediately. Moreover, after finishing the work in the radiochemical laboratory, it is also necessary to ensure that the personnel is not carrying with him some radioactive isotopes due to some contamination, because this will not only be dangerous to the person, but also to the society as well. In addition to this, while working with γ-source or neutron source, it is also necessary to get the information about the dose being received. For this purpose, we need a device which can give an instantaneous dose in terms of rad/hr received by the person. Finally, there is also a need to have some knowledge of the accumulated dose due to various radiations received by the personnel working in the laboratory over a period of week/months/year. For these purposes, some special type of monitors are used. These are being discussed in this section.

12.7.1   Portable Hand Monitor

The portable hand monitor usually consists of an ionization chamber designed to have a response, which is approximately proportional to the dose in rad/hr. The portable monitor operates usually with a battery operated valve voltmeter. This monitor is used to measure the dose-rate during the dispensing of radioactive materials.

12.7.2   Pocket Dosimeter

A pocket dosimeter consists of a chamber containing a metalized quartz fiber, which is charged to a fixed potential. Any radiation passing through it causes ionization, which allows the charge to leak away. The movement of the fiber indicates the integrated dose. The greatest advantage of these chambers is that they can be read daily or even for a short period. The scale gives the dose-rate in rads. This instrument is often used while dispensing a radioactive material.

12.7.3   Photographic Film Badge

A photographic film badge consists of a small film enclosed in a paper, in a holder. The accumulated dose in a given period can be estimated by comparing the exposure of the film with a standard film. Each individual working in the radioactive laboratory should carry such a film badge. Persons working with softer radiations, such as β-particles, should wear a special type of film badge, which is sensitive to weak radiations.

12.7.4   Monitors for Survey Work

A common form of survey monitor is the Geiger counter, connected with a long coaxial cable to a rate-meter plus a speaker. This type of counter is designed to give the average count rate on the scale, as well as make a sound, according to the amount of radioactive material near the counter. This provides a quick method for monitoring the laboratory and apparatus, as well as for indicating the background level of activity present in the laboratory. Such type of monitors are usually kept near the entrance of a radioactive laboratory, so that level of activity can be immediately noticed. This is very useful, especially if someone contaminated with radioactive isotopes passes by the counter, the level of sound suddenly increases, which warns the person of some contamination.

12.8   Design of a Radioactive Tracer Suit

A radioactive laboratory should possess all the basic features of a chemical laboratory, but due to the hazards associated with the radioactive materials, while designing the radioactive laboratory some of the following care must be taken.

1. The activity must be confined to a particular area and the level of activity should increase in a specific order.
2. The counting rooms should be situated in the area of minimum background. For a low level work, a special counting room should be used and be treated as an inactive area.
3. Outside the active area, there should be a provision for calculating the results, or for readings etc.

4. There should be separate rooms for the storage of radioactive materials and for dispensing highly active materials. These areas should be treated as the most active ones.

5. The research laboratory, or training laboratory, should be situated near to the dispensing rooms, so that the active material may be transferred without being carried for a long distance.

6. In the research and the training laboratory, there should be a provision for a big fume cupboard, so that experiments with volatile radioactive materials may be carried out in it without any difficulty.

7. At the entrance to the laboratory, there should be provisions for washing and pegs for laboratory coats etc.

8. In buildings where air-conditioning and fume cupboards are used, the arrangement should be made so that air from the active area may not be sucked to the inactive area.

9. The flow of water into the building should also be such that the water may not flow back to the main building from the active laboratory, especially from the dispensing room.

10. There should also be a provision for a space where inactive work such as the separation of certain compounds etc., can be carried out.

11. A room for a safety officer, inactive chemical storeroom, lecture room, toilets etc., and other necessary rooms should be built along with the radiochemistry suite, and these should be treated as an inactive area.

12. In a tracer laboratory, it is necessary to appoint a safety officer, whose duty is to maintain discipline in the laboratory and to keep a record of individual doses received. He/She should also have arrangements with the appropriate medical authority, who can be called in case of an emergency. He/She should have a complete record of various isotopes which are being used by the workers.

13. The interior design of the radioactive suite is also important. All the water and the gas pipes etc., must be boxed in to assist in cleaning the laboratory. Benches should be polished and covered with bituminous paper, so that any spill of radioactive material may be cleared up easily. The floor must be covered with linoleum, which is kept well waxed, otherwise it may absorb active solution and decontamination may be impossible.

14. The neutron source room should be doubly shielded with thick concrete walls from each side, and when convenient the neutron source may be kept underground, the earth acting as shielding.

15. In the radiochemistry suite, it is preferable to have two doors (not on the same wall), so that in case of an accident, the workers can escape easily.

Considering all these factors, a self-explanatory simple design for a radiochemistry suit is shown in Fig. 12.1. In this drawing, the activity increases as one goes from the entrance toward the neutron source.

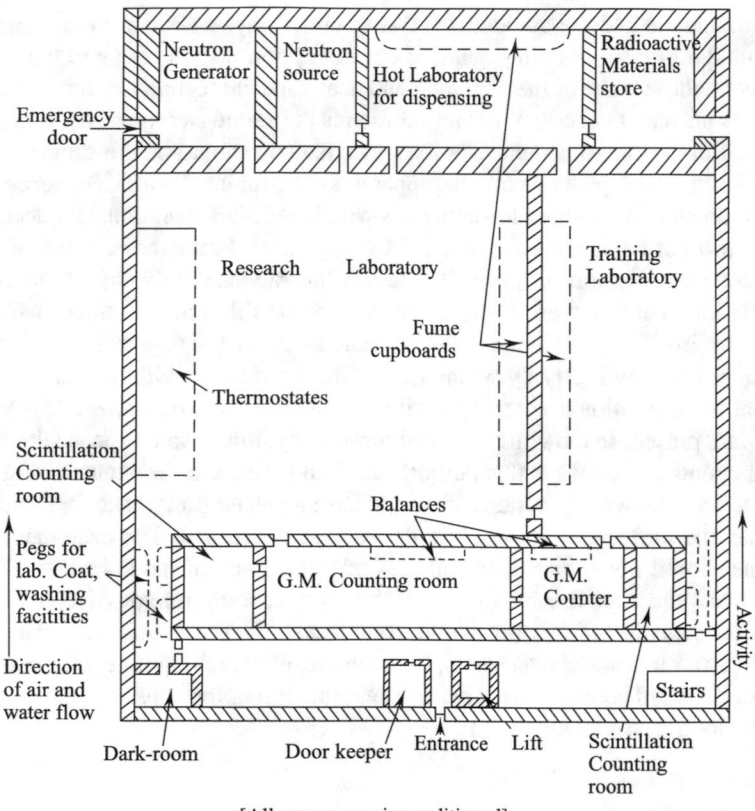

Fig. 12.1 A schematic layout of a typical radiochemical training/research laboratory

12.9 Decontamination of Apparatus

In a chemical laboratory, cleaning is done at the microlevel of contamination. In the radiochemical laboratory, the cleaning of the apparatus has to be done at the atomic level of radiation so that the level of activity present in the apparatus after cleaning is very near the background count. For radiochemical work, one may be able to tolerate chemical contamination but contamination with radioactive isotope cannot be tolerated at all. If the level of activity due to contamination of radioactive isotopes is higher than the background count, the experiment for which the apparatus is to be used will give erroneous results. Therefore, the decontamination of the apparatus is very important. Even a trace of radioactive material present in the apparatus (especially if it is of high specific activity i.e., activity per unit of amount) may be sufficient to ruin the result completely. The decontamination of apparatus in most cases is easy, but with isotopes like bromine, which sticks on to the glass very firmly, decontamination becomes a problem.

Contamination of the apparatus is due to either physical adsorption, which can be washed off by any wetting agent, or chemisorption, which is due to the residual valency on the surface of the container and needs special techniques for decontamination. A universal decontaminating solution is the stable isotope of the radioactive isotope. But the chemical nature of the stable isotope should be the same as that of the radioactive isotope. For example, apparatus contaminated with ^{82}Br can be easily decontaminated by washing the apparatus with 0.1M NaBr solution. But decontamination will not be to the satisfaction if the apparatus, in this case, is washed with 0.1M aqueous solution of bromine, because the radioactive isotope is present as bromide ion. This method is known as the isotope dilution technique, where the radioactive isotope gets diluted with the stable isotopes and thus its level of activity get reduced tremendously by washings; two to three times with the stable isotope solution. Such dilution can occur only if both, the radioactive isotope and the stable isotope are present in the same chemical form. Some times, some special decontaminating agents are used for this purpose. In such cases, the decontaminating agent must have a good wetting action and should have a greater preference for the absorption than the radioactive material on the surface concerned. The decontaminating reagents should not spoil the smoothness of the surface of the container, because it will make the decontamination very difficult e.g., sodium hydroxide or chromic acid. The reagents which form soluble complexes with the radioactive material are also preferred in some special cases, for example, ethylene-diamine-tetraacetic acid (E.D.T.A). Detail procedures for the decontamination of the apparatus are discussed in the following sections.

12.9.1   Process of Decontamination

It is customary to wash the apparatus with a carrier solution which dilutes the activity to a considerable extent. The carrier solution is the aqueous solution of the stable isotope of a radioactive isotope, which is to be decontaminated. The chemical nature of both of them should be in the same form. In addition, dilute acids, such as nitric acid, inhibited orthophosphoric acid may also be tried. The complexing agents such as EDTA or any suitable ion-exchange resins may perhaps be useful for this purpose.

Hands can usually be decontaminated by washing with soap and water. Sometimes titanium dioxide paste or EDTA solutions are useful. A good general decontamination agent is a saturated potassium permanganate solution, followed by 5% sodium bisulphite solution. But chemical washings should not be employed regularly, since the skin being porous may lead to penetration of the activity into the body and thus it may cause an internal hazard. In the case of internal contamination due to any cuts etc., bleeding should be encouraged and proper medical attention should be given. Bleeding helps in eliminating the radioactive nuclide.

12.9.2 Disposal of Radioactive Waste

Unlike chemical laboratory, where chemicals and washings coming out of the apparatus are normally thrown into the drainage, in the radiochemical laboratory, a very special care has to be taken for disposing any radioactive chemicals, because it might cause danger to the public, as well as to the personnel working in the building of the radiochemical laboratory. Therefore, safe disposal of radioactive waste material arising as a result of the experimental work in a tracer laboratory should be considered before such work commences. The disposal of active material must also be done with the consent of local Medical Officer (Safety Officer) of Health and a record of it should be kept for checking. However, the following precautions must be adhered to while disposing the radioactive waste:

1. Short-lived materials should be stored in some unfrequented place until their activities have decayed to a sufficiently low level for disposal as an inactive waste.
2. The long-lived radioactive isotopes, especially highly toxic isotopes, such as Strontium-90, or α-emitter isotopes are concentrated and sent to the Atomic Energy Establishment. However, moderately toxic or slightly toxic isotopes are disposed into water drainage, provided the dose-rate at the surface of the tap is not above the maximum permissible level.
3. It is preferable to dilute carrier free materials of high specific activity with a non-radioactive carrier (which should be in the same chemical form as the active material). This decreases the specific activity and reduces the health hazard. Radioactive diluted liquids should be poured into those sinks which are directly connected to the main drainage system of the building by closed drains. The dose-rate at the surface of the tap should not exceed 10 mrad/8hr a day.
4. In the tracer laboratory, gas disposal is not normally necessary. However, the gases coming out of the fume cupboards are filtered and filtered gases are disposed to the atmosphere. It is preferable to monitor the filter periodically in order to check radiation level due to any possible radioactive dusts.

12.10   Discipline in the Radioactive Laboratory

The object of discipline in the radioactive laboratory is to prevent radioactive contamination and to safeguard the health of the worker. In each radioactive laboratory certain rules are set up by the Safety Officer, and the workers must obey them.

The following are a few brief outlines of such rules:

1. A laboratory coat or apron with a film badge should always be worn while in the active area and must be taken off when coming to the inactive area. This rule is made to minimize contamination.
2. Rubber gloves should be worn when performing chemical manipulations with radioactive materials. If there is any cut on the hand, the rubber gloves must

be used or whenever possible, plastic tape and gloves should be taken off in such a way that unprotected fingers do not touch the outside of the glove. It is better to wash the gloves before they are taken off, and then monitored, or at least a periodical washing should be done. It is noteworthy that the gloves do not protect from radiation hazard (except for α-particles), but are used to prevent the contamination of the hands, in case of any spill of radioactive materials.

3. Mouth operations, such as eating, drinking, smoking or even pipetting by mouth etc., are strictly forbidden, in order to avoid the internal hazards. When leaving the radioactive area, hands must be washed and monitored.

4. It is preferable to carry out radioactive experiments in shallow trays, so that laboratory furniture does not get contaminated in the event of a spill or breakage of apparatus.

5. Work with a volatile compound must be carried out in the fume cupboard.

6. The radioactive wastes, either liquid or solid, should be kept in a special container.

7. The laboratory should be periodically monitored against contamination.

8. While carrying out any new experiment with radioactive isotopes, a dummy experiment with non-radioactive material must be performed so that necessary precautions to avoid any accident can be thought off.

9. In the laboratory, there must be a Geiger counter monitor connected with a loudspeaker in order to regularly hear the background level.

Summary

Handling of radioactive isotopes needs the worker to be aware of the health hazards associated with radioactive isotopes. There is a need to take care of the disposal and contamination problems associated with some of the hazardous radioactive isotopes like ^{137}Cs. Safety of the personnel working in the radiochemical laboratory is most essential. It is, therefore, necessary to understand the reasons, which cause the health hazard. In this chapter, therefore, we studied these aspects and discussed to explain the various factors necessary to be considered before launching to work with radioactive isotopes. The design and necessary requirements of a radiochemical laboratory are also discussed.

Chapter 13
Radiochemical Separation Techniques

13.1 Introduction

In this chapter, efforts are made to give a brief outline of a few techniques, which are routinely followed in radiochemical experiments. It is important to realize that in radiochemical separation techniques, emphasis is given to the purity of radioactive material and not much to their chemical purity. This is mainly because the amount of activity is measured with the help of the various radiations emitted by the radioactive isotopes. Therefore, in radiochemical separation techniques, we are concerned with the efficiency of separation at the atomic level and hence the technique adopted must be very specific to the requirements. The radiochemical separation techniques normally are used for the separation of radioactive isotopes from the target material used for carrying out nuclear reaction initiated by neutrons or any heavy ions accelerated particles. Sometimes one needs to separate the radioactive isotopes from some chemical reactions being conducted or biological processes to study the nature of chemical reactions or biological reaction processes etc.

In general radiochemical techniques provide the means for the following:

1. Separation of radioactive isotopes from the original target material (i.e., the material used for generating radioactive isotopes) or from any other unwanted radioactive isotopes which have been produced during nuclear irradiation.
2. Preparation of the radioactive source for counting purposes and
3. Identification of various radioactive isotopes produced during nuclear irradiation.

Some of the techniques used for these purposes are discussed here.

13.2   Separation and Purification Techniques

The basic methods used for the purification of radioactive samples are those of analytical chemistry. These analytical methods may have to be modified or carried out with care in order to achieve maximum radiochemical purity, and it may often be less harmful to have a few milligrams of inert impurity in the final sample than 10^{-9} g of a radioactive contamination. The speed of the separation technique in some cases may be more important than high chemical yield or even great chemical purity. A very good and high yield separation scheme, which requires an hour to perform, is of little use for isolating and studying a nuclide with a half-life of a few minutes. Also, the method preferably should not lead to high contamination of apparatus; this may cause serious losses of active material and give much trouble in decontaminating the apparatus for further use. Considering these stringent requirements, few specific techniques are discussed here.

13.3   Co-precipitation

The amount of radioactive nuclides formed in an irradiated target material is often very low to permit the solubility product of an insoluble compound to be exceeded. The addition of a chemically identical element (a carrier) in the same chemical form, allows the solubility product to be exceeded and results in precipitation. An advantage of this technique is that the amount of material associated with the activity can be very closely controlled by the addition of the known amount of the carrier.

13.3.1   Carriers for the Separation

In nuclear reactions, other than (n, γ), the daughter isotope is always a different chemical species than the parent. Hence, the number of daughter isotope formed may be of the order of 100–10,000 atoms. Moreover, chemical separation needs some glassware to carry out the experiment. The radioactive isotope formed during the nuclear reaction can thus very easily be adsorbed on the surface of glassware, hence there can be loss of radioactive isotope during chemical maneuvering. In addition, these atoms may be required to be precipitated or solvent extracted for the separation. The amount of these radioactive species would be very low to be visible in precipitation or other techniques. Therefore, the chances of losing the radioactive atoms become great. However, if after nuclear reactions, the substance is mixed with an inactive substance (known as **carrier**) which has the same chemical property as the tracer, then the radioactive isotope produced in a nuclear reaction is diluted (i.e., its specific activity is lowered) and loss due to adsorption etc., during the chemical separation is minimized.

Carriers can be of three types:

1. Isotopic carrier,
2. Non-isotopic carrier and
3. Hold-back carrier.

13.3.1.1 Isotopic Carrier

Isotopic carriers are those, which should be of the same chemical element as that of the radioactive substance but should be a stable isotope of the tracer. For example, iodide ions can be used as an isotopic carrier for ^{131}I-ions. The disadvantage of this type of carrier is that it lowers the specific activity of the material, because the radioactive species cannot subsequently be chemically separated from the carrier. Hence, when a thin, virtually weightless deposit of the material is required for counting; the use of isotopic carriers should be avoided. Moreover, a stable isotope of iodine cannot be used as a carrier for ^{131}I-ions, because though both belongs to the same element, but their valences are different. In such cases, it is preferred to carry out at least a couple of oxidation and reduction processes to make sure that both the carrier and the radioactive isotope are in the same chemical form.

13.3.1.2 Non-isotopic Carrier

A "non-isotopic" carrier is an element, which has similar chemical properties as the radioactive isotope, but is not an isotope of the tracer. A non-isotopic carrier is sometimes used where there is no stable isotope available (e.g., barium is used for radium isotope separation) or when it is desirable that the carrier should be subsequently separated from the radioactive nuclide, especially when the original specific activity is to be recovered. For example, the separation of Protoactinium-233 from neutron irradiated thorium nitrate solution is carried out by adding manganese nitrate and precipitating manganese dioxide from the solution by the addition of potassium permanganate. The advantage of this carrier is that the specific activity of the tracer does not change, provided at the end of chemical separation, added inactive carrier could be removed from the tracer.

Another type of non-isotopic carrier is called "**scavenger**". Such carriers are mainly gelatinous precipitates, such as ferric hydroxide or aluminum hydroxide. Since freshly prepared hydroxides have a large surface area, adsorption of radioactive isotopes present in the solution may remove the nuclides present in trace amounts in the mixture.

13.3.1.3 Hold-Back Carrier

Hold-back carrier can be either isotopic or non-isotopic type. Coprecipitation of unwanted radioactive species during the precipitation is sometimes reduced by diluting them with a stable isotope of the same element, in the same chemical form. For example, precipitation of radioactive isotope ^{131}I while precipitating ^{35}S as barium sulfate from a mixture containing ^{131}I- and ^{35}S-ions is prevented by adding 0.1M KI into the mixture before carrying out the precipitation. Potassium iodide holds back ^{131}I in the solution while $Ba^{35}SO_4$ is precipitated. Such type of carriers are called as "hold-back" carriers.

A combination of the hold-back carrier with a scavenger works as an artificial filter paper that can remove micro-concentration of impurities, which have no carrier, from a micro-component. However, although the radiochemical purity of the nuclide is considerably increased by this procedure, its specific activity is lowered.

13.3.2 Condition for Effective Use of Carrier

Carrier can only be effective if the chemical nature of the carrier is the same as that of the tracer. This applies to both the isotopic and non-isotopic carriers. For example, iodide ions can be exchanged for iodide ions only and not for iodate ions. Hence, if there is uncertainty in the chemical state of the tracer present in the sample, it is always better to add the carrier irrespective of its valence state and then carry out three or four times the oxidation and reduction processes of the sample after adding the carrier, so that all the substance with tracer come back to its one chemical state.

13.4 Solvent Extraction

In solvent extraction, the substance to be separated is mixed with two immiscible solvents (normally organic solvent with aqueous solvent) and shaken vigorously for some time and then two layers are allowed to settle. The radioactive isotope gets distributed between the two solvents. Since the distribution coefficient for the radioactive isotope with two solvents is different, one of the solvents will get concentrated with the radioactive isotope leaving behind other constituents in other solvents. Seaborg and Graham have shown that the micro-components distribute themselves between the two immiscible solvents with the same value of partition coefficient as was present in macro-amounts. Hence, the separation by this technique can be carried out without any use of the carrier. This method can be applied to almost all soluble substances, organic or inorganic, and can be made quantitative and often remarkably specific. Solvent extraction methods are useful in radiochemical separations because of the ease and speed of manipulation.

The ionic compounds are not soluble in covalent organic liquids; the solvent extraction of metallic ions, therefore, requires the presence of some substances, which convert them into neutral soluble complexes. The use of organic complexing agents such as 8-hydroxyquinoline, dimethylglyoxime etc., are in common use for this purpose. This technique has been used extensively in radiochemical separation. For example, separation of uranium and plutonium from the fission products in reactor fuel elements has been successfully achieved by dissolving the fuel elements in nitric acid, from which uranium and plutonium are separated by extraction from a variety of solvents, such as tributylphosphate diluted with an inert solvent like kerosene. Plutonium (in trivalent state) is extracted by washing the solvent phase with an aqueous phase containing reducing agents such as ferrous ions, whereas uranium is recovered by washing the organic phase with water.

13.5 Electrochemical Methods

The electrochemical method is nowadays used extensively because the contamination by adsorption of micro-components from the solution is very small, moreover, it is a very rapid method. The electrolysis may be carried out with or without the use of a carrier. For example, the deposition of Silver-110 with or without the presence of natural silver as a carrier, on a gold disk as a cathode, using ammonical solution, can give about 99.9% yield. Thin sources, showing very little self-absorption, can be prepared by this method.

The electrolysis can be performed in two ways. In the first, the metal can be deposited on the cathode from the appropriate solution of its ions; for example, a protoactinium source can be prepared by electrodeposition on various metal cathodes from a very dilute and slightly acidic fluoride solution. In the second method, the species in the region of high pH, produced by the discharge of hydrogen ions at a cathode, is precipitated in an insoluble form on to the cathode; examples are lanthanides and actinides precipitated by hydroxide.

13.6 Volatilization and Distillation

Volatilization or distillation is often a quick method of separation provided the radioactive materials to be separated have widely different vapor pressures. For example, Iodine-131 may be volatilized from a neutron irradiated samples of tellurium oxide.

13.7   Chromatography

The term **chromatography** is commonly applied to the separation of substances by selective distribution between a flowing fluid and an insoluble solid known as the support. Chromatographic technique may be used for processing micro amounts of material for preparative purposes or micro amounts for assay purposes. Chromatographic techniques for the purpose of radiochemical separations can be classified as:

1. Partition chromatography—Paper chromatography
2. Ion-exchange chromatography
3. Adsorption chromatography—Column chromatography.

13.7.1   Paper Chromatography

Paper chromatography is broadly applicable for the qualitative and quantitative analysis of mixtures. This is based on the partition of the substance to be separated between a fixed and mobile phase, and the separation depends on the different distribution coefficients.

In this technique, the solution which is to be analyzed or separated is placed on a sheet of filter paper in the form of a small spot near one end (Fig. 13.1A). After the spot "*A*" has been dried, it is hanged vertically in a gas jar covered with a lid, containing a suitable solvent (Fig. 13.1B). The solvent is allowed to move vertically through the paper, starting at the end of the spot. The various components of the original mixture travel up with the solvent, but at different rates. After the solvent has traveled a sufficient distance, the paper is removed and dried by an infrared lamp. The situation of the constituents is found by the various methods, such as scanning the papergram with a G.M. counter (Fig. 7.5); by exposing the *X*-ray film to the papergram and then

Fig. 13.1 **A** is a filter paper over which a radioactive spot is kept and is hanged vertically in a gas jar **B** containing a solvent and **C** filter paper after the solvent has traveled to the top of the paper. In order to calculate the R_f factor, a fresh spot of radioactive material is put at its original spot and a line is drawn with radioactive liquid to find out the solvent front

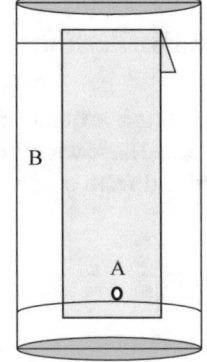

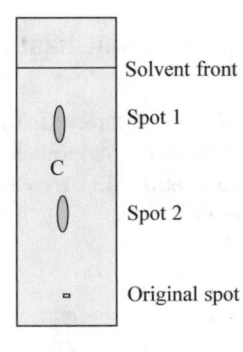

developing it (Fig. 7.6A and B) or by cutting the papergram into small strips, finding the count rate of each strip and drawing a histogram. The separated constituents (Fig. 12.1C) are then removed from the papergram by extracting with the help of a suitable solvent. In order to confirm the identity of the isotope, it is customary to calculate the R_f factor for each separated nuclide. R_f is defined as the ratio of distance traveled by the radioactive isotope spot and the distance traveled by the solvent front.

The R_f factor of an ion has a characteristic value under the given conditions, and may give a preliminary indication of the ion's identity. For example, Fig. 7.6A and B were obtained by scanning the paper chromatogram with a G.M. counter and auto-radiograph (positive of an X-ray exposed film) respectively, of a chromatogram of a reactor irradiated mixture of sodium bromide and potassium fluoride; hydrochloric acid methanol was used as solvent. Figure 7.6B was obtained from the auto-radiograph produced after the radiation had penetrated 11 layers of X-ray film (Ilford industrial G), and indicates the relative penetrating power of radiations. R_f values calculated from these figures are shown in the table:

| Nuclides - energy | | R_f factors | |
Nuclide	E_{max} (MeV)	Scanning	X-ray film
Potassium-42	3.6	0.16	0.16
Sodium-24	1.39	0.29	0.24
Bromine-80	1.99	0.81	0.83

13.7.1.1 Factors Influencing the Separation of the Species

1. *The R_f factors* of certain nuclides may be small or very close to the values for other species present in the mixture; this makes a satisfactory separation difficult. In cases like this, the multiple development technique may be helpful. A developed strip is dried and rechromatographed in the same direction with the same solvent; this allows more time for the spot of low mobility to move or spots of closely similar R_f values to separate. A zone containing two or more unseparated substances may then be cut out from the rest of the chromatogram, sewed to a second strip, rechromatographed with a different solvent. However, the method suffers from the disadvantage of requiring a longer time, and might not be useful for short-lived radioactive nuclides.

2. *The amount of the sample* should not be sufficient to overload the paper, because this leads to a tailing effect and bad separation of the nuclides. For a large amount of sample, thin layer chromatography (TLC) is preferred. For TLC on a glass sheet, a thin layer of fine alumina or silica gel is prepared by spreading a thin aqueous slurry of alumina or silica gel. This is then used in a similar fashion as the paper chromatography. The separated spot can be removed from the glass sheet for further analysis. Carrier free samples, however, may not move so readily on the paper as it can with a thin layer of silica gel.

3. *The selection of the paper* is also important and some factors such as the presence of impurities, the wet strength, the thickness, the flow rate of solvent etc., are important.

4. *Paper chromatograms should* not be dried by blowing with air, because this may cause some of the radioactive nuclides to blow away from the paper. With a β-emitter the paper must be thoroughly dried, otherwise the residual solvent will absorb some of the radiation; this is important when the paper is going to be counted directly. However, each separated constituent may be extracted from the paper and can then be counted.

5. *Many radioactive nuclides* produced by a nuclear reaction have a half-life such that they must be separated and identified quickly. Paper chromatography can in such cases be carried out rather rapidly by choosing a fast-moving solvent and a short paper strip. Usually, a less viscous liquid moves faster and descending chromatography gives a faster flow rate than ascending.

13.7.2 Paper Electrophoresis

In paper chromatography, the solvent and ions present in the mixture moves due to gravitational force, but in paper electrophoresis, ions are forced to move by application of potential difference across a paper strip. The paper electrophoresis has been in use since 1948. Electromigration depends on the different velocities of the substances under an electric field, so that in principle the stationary solid phases (paper) serves merely to prevent mixing by convection, while an associated liquid phase serves to keep the migrant in a mobile form with a suitable charge.

In this technique, unlike that of paper chromatography, a filter paper is moistened with the chosen so-called background electrolyte solution (such as dilute sulfuric acid, sodium sulfate etc.). Each end of the strip is dipped into the electrolyte and a potential difference is applied. Unlike the paper chromatogram, the spot of the sample is placed in the center of the paper (length-wise) and when the potential is applied, the ions migrate at different speeds and in different directions depending upon their nature of the charge. Separation is thus achieved. The separated spots are spotted on the paper by adopting any of the techniques described earlier. The separated spots are extracted from the paper by a suitable solvent.

This method is usually more rapid than that of ordinary paper chromatography, but the calculation of a separation factor, such as R_f values, is not so easy because there may be many factors affecting the separation. Therefore, for identification of the separated spots, the half-life or energy of radiation emitted by the isotope will have to be determined.

13.7.3   Ion-Exchange Method

An ion-exchange resin consists of a polymeric framework of an insoluble hydro-carbon to which ionizable groups are attached. The resin may be visualized as an insoluble polyelectrolyte. These polymers are obtained by polymerization of phenol and formaldehyde or by copolymerization of styrene and divinylbenzene. Specific ionizable groups are introduced to provide the desired exchange properties. In simpler words, ion-exchange substances are insoluble solids with acidic or basic properties, capable of forming loosely bound compounds with cations or anions respectively. These ions exchange for an equivalent amount of other ions of similar charge i.e., when the ion-exchanger comes into contact with electrolyte solutions.

Ion-exchange resin can be classified as cation-exchange resin or anion-exchange resin, depending upon the nature of the ionizable or exchangeable group. These may be further classified as a strong or weak acid or base, according to the characteristic of the ionizable group. The strong functional groups are well suited for the inorganic separation and the weak groups for organic separation e.g., mixture of amino acids.

Through the process of adsorption and diffusion, an ion in solution in contact with an ion-exchange resin can exchange with or replace an ion supplied by the functional or "ion-active" group of the resin. Thus, for a cation-exchange resin, the functional group in the acid form, such as the sulfuric acid (-SO_3H) and carboxylic acid group (-COOH) can ionize and reversibly exchange for H^+ for cations such as Na^+, Ca^{++}, La^{3+}, Th^{4+} etc., in an appropriate stoichiometric ratio. The same functional group in the salt form, e.g., -SO_3Na and -COONa, can undergo exchange with H^+ or other metal cations.

For anion exchange resin, quaternary ammonium bases (=N—OH,) and amines (−NH_2 > NH, =N correct this) are typical strong and weak functional groups, respectively. The strong quaternary ammonium base type resin exchanges rapidly with anions, e.g., Cl^-, $FeCl_4^-$ in the base or the salt form.

An ion-exchange column can be used to separate not only one constituent of a sample, but many constituents, provided they have different affinities for the ion-exchanger. The exchange capacity depends upon the type of polymer and the type of functional group. The capacity of a weak acid or weak base type is dependent upon the pH of the contacting solution. For strong functional groups, the capacity is independent of pH except for effect a resin from weak groups which may also be present in the resin to a minor extent. At very high acidity the capacity of a cation-exchange resin is decreased for ions of low affinity. Ion-exchange capacity may be expressed in several ways, milliequivalent/g of dry resin or milliequivalent per milliliter for wet resin.

Ion-exchange resins may be used in the separation of ionic species present in a sample. The separation is carried out either by contacting the solution containing the ionic material with the resin in a batchwise mixing process or by passing the solution through a bed of resin contained in an ion-exchange column (Fig. 13.2A). The former technique is used for equilibrium distribution studies; but the non-equilibrium column

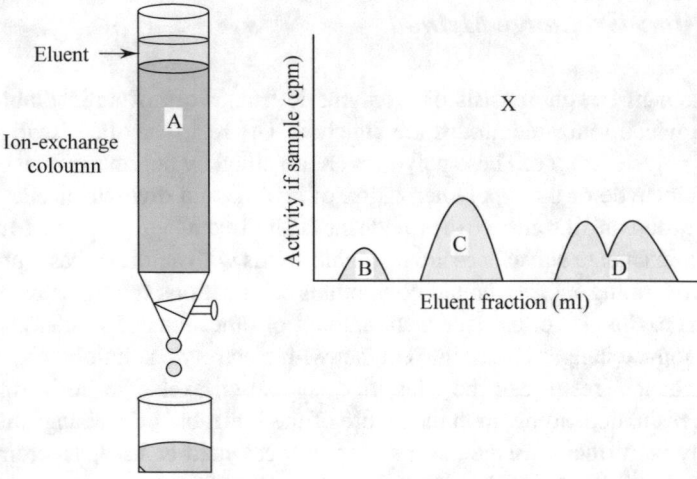

Fig. 13.2 A A schematic diagram showing the arrangement of the ion-exchange column and the activity present in the different fraction of eluent collected from the column **B** and **C** shows the distribution and their respective magnitude of activity recorded by two well separated fractions and **D** shows that two types of isotopes which could not get well separated

technique is generally used for actual separation. In the former technique, when a solution of ionic species is passed through the column; ions migrate down the column at a rate dependent upon their exchange affinities. The first ion to appear in the effluent is the one most weakly adsorbed (Fig. 13.2B). The other ions appear in order of increasing adsorption (Fig. 13.2C).

An ion-exchange column is prepared by making a slurry of an ion-exchange resin in water and pouring it into a glass column containing glass wool at the bottom. The solution to be exchanged is then poured into the top of the column. The ions are first adsorbed from a dilute solution in a narrow band at the top of the column. The exchange occurs in the column and on developing the column with a suitable solvent, the exchanged ions, now in the resin, can be recovered and then removed by the addition of an eluting agent or an elutriate which causes the band of adsorbed ions to separate, as it moves down the column, into a number of bands corresponding to the number of ionic species being separated. Each drop, or each half ml. coming out of the column is collected separately and its activity is measured, The degree of separation depends upon the nature of the eluting agent as well as the resin.

The eluent thus collected is used to measure its activity and a graph is plotted as a function of the volume of the effluent or milliequivalents of ion passed through the column and the activity recorded for each fraction of eluent (Fig. 13.2A). The nature of the distribution of activity can be asymmetric sigmoidal shape (Fig. 13.2B and C) or may overlap in the case of poor separation (Fig. 13.2D).

Complexing agents such as ammonium citrate, ammonium lactate etc., are also used to increase the separating power of an ion-exchanger where it has similar affinity for more than one ion, as the case represented in Fig. 13.1D. Such agents are widely used for the separation of lanthanides and actinide elements, whose chemical properties are very similar. For example, Choppin and his co-workers separated the transuranic elements from sameracium, by the ion-exchange method. They used Dowerex-50 cation-exchanger at an elevated temperature (80 °C). A constant temperature was maintained by enclosing the ion-exchange column in the vapors of trichloroethylene (B.P. 80 °C). Ammonium lactate complexing solution was used to elute the ion-exchanger and each drop of the elute was collected separately, a source was prepared from each and counted. A histogram of activity present in each drop eluted from the column versus the number of drops removed from the column showed that the elements had been separated and came from the column in reverse order of their atomic number.

13.7.4 Nuclear Recoil-Method

It is sometimes possible to separate the products of the nuclear reaction by making use of the nuclear recoil phenomenon. The nuclei of the recoiling fragments may travel rather a long way in the target material and may leave it altogether if the target material is less than the range of the energetic fragments. For example, Wolfgang separated fission products from Neptunium-239 and Uranium-239, using U_3O_8 (particle size one micron) as target material. The fine particles of U_3O_8 were prepared in gelatinous suspension and were irradiated in a neutron source (nuclear reactor). The gelatin was then liquefied and the mixture was centrifuged. The supernatant liquid contained the fission products, whereas Neptunium-239 remained in the uranium oxide. The separation gave a carrier-free Neptunium-239, free from fission products, after the decay of Uranium-239 ($T_{0.5}$ 23.5 min.).

13.8 Activation Analysis

In chemical analysis, the degree of accuracy depends upon the method of analysis. In cases where chemical method of separation is not available; usually activation analysis is tried. One of the important applications of nuclear reactions in analytical chemistry is also the **activation analysis**. This method is based on the principle that radioactive nuclides are formed by the interaction of a nuclear particle with an element which has an appreciable ability for the reaction to occur. This ability is measured in terms of **cross-section** for the nuclear reaction. The induced activity due to nuclear reaction is then detected and measured to determine the amount of element present in the sample material to be analyzed. This type of analysis is basically useful because of the following factors:

1. Each radioactive isotope emits a characteristic radiation and its energy can be taken as evidence to identify the nuclide.

2. One can measure the activity and hence its energy, or half-life of the isotope without performing any chemical separation.
3. One can analyze the content of the material without destroying the form of the sample.

13.8.1 Theory of the Activation Analysis Technique

Whenever a sample is put into a reactor or neutron source, the neutron is absorbed by the sample to produce a radioactive isotope. Let's assume that the activity (A_t) is present for a given isotope of an element having weight $w_A g$ of atom A after times "t" of the start of the irradiation and atom B is produced due to the nuclear reaction which in turn decays to atom C with a decay constant of λ_B.

$$A(n, \gamma)B \xrightarrow{\lambda_B} C$$

Rate of production of atom B per unit time

$$n \times \frac{N \times w_A}{M} \times \phi \times \sigma \times A$$

where

$n =$ number of neutron/cm^2/s,
$\phi =$ isotopic abundance,
$\sigma =$ cross-section,
$A =$ number of atom A,
N and M are the Avogadro number (6.023×10^{23}) and the molecular weight of atom A respectively.

Rate of decay of B atom/unit time

$$\frac{dN_B}{dt} = N_B \lambda_B$$

Therefore, the rate at which the activity of B will accumulate during the neutron irradiation is given by

$$\frac{dN_B}{dt} = \frac{N w_A n \phi \sigma}{M} - N_B \lambda_B. \tag{13.1}$$

Alternatively, the activity of B present at any given time "t" can also be expressed as

$$N_B = \left(h_A + h_B e^{-\lambda_B t}\right) \tag{13.2}$$

where h_A and h_B are some unknown constants. Differentiating this equation gives us

$$\frac{dN_B}{dt} = -h_B e^{-\lambda_B t} \lambda_B \tag{13.3}$$

substituting this value into Eq. (13.1), we have

$$-h_B \lambda_B e^{-\lambda_B t} = \frac{N w_A n \phi \sigma}{M} - N_B \lambda_B. \tag{13.4}$$

Thus

$$-h_B \lambda_B e^{-\lambda_B t} = \frac{N w_A n \phi \sigma}{M} - (h_A + h_B e^{-\lambda_B t}) \lambda_B. \tag{13.5}$$

or

$$-h_B \lambda_B e^{-\lambda_B t} = \frac{N w_A n \phi \sigma}{M} - h_A \lambda_B - h_B \lambda_B e^{-\lambda_B t} \tag{13.6}$$

hence

$$\frac{N w_A n \phi \sigma}{M} = h_A \lambda_B \tag{13.7}$$

or

$$h_A = \frac{N w_A n \phi \sigma}{M \lambda_B} \tag{13.8}$$

Now using Eq. (13.1)

$$N_B = \left(h_A + h_B e^{-\lambda_B t} \right)$$

and solving it for time $t = 0$, we have $N_B = 0$. Thus, $h_A = -h_B$. Substituting this condition into the Eq. (13.2), we have

$$N_B = \frac{N w_A n \sigma \phi}{M \lambda_B} - \frac{N w_A n \sigma \phi}{M \lambda_B t} e^{-\lambda_B t} \tag{13.9}$$

i.e.

$$N_B \lambda_B = \frac{N w_A n \sigma \phi}{M} (e^{-\lambda_B t}) = \text{activity of atom } B \text{ per unit time} \tag{13.10}$$

Thus, by measuring the activity of the irradiated sample (if it contains one type of atoms) or of the sample chemically separated, the amount of atom A can be determined from this equation, because all other terms are known.

Sometimes, the nuclear data for the isotope or the exact neutron flux is not easily available. For such cases, along with the sample, a standard sample (i.e., an element which is expected to be present in the unknown sample) are irradiated together and the amount of the sought element in the sample is calculated from the following equation:

$$W_1 = \frac{X}{Y} \times W_2 \qquad (13.11)$$

where W_1 is the weight of the sought element in an unknown sample, W_2 is the weight of the sought element in the standard sample, X is the activity of the sought element in an unknown sample and Y is activity of the sought element in the standard sample. If there be a need to carry out any chemical separation technique of irradiated sample, then the standard sample should also follow the similar chemical treatment. The advantage of this method is that if there is any loss of activity due to chemical separation, both samples will experience the same loss, and hence can be nullified in calculating the exact weight W_1.

13.8.2   Experimental Procedures

The sample to be analyzed is irradiated in a nuclear reactor or in any other particle generator, for a required period. A reference sample is also irradiated in exactly the same way. The measurement of activity can be carried out in two ways, depending upon the type of radiation emitted and the number of radioactive impurities present in the sample. For example, if the radioactive nuclide produced is a γ-emitter, and there are no other impurities which give γ-radiation, then a direct measurement of the sample's total γ-activity could give an idea of the amount of the required element present in the sample.

On the other hand, if there are several radioactive nuclides present or if there is no idea of possible impurities present in the sample, then a chemical separation is preferred. Thus the first step in measuring the activity or activities in an irradiated sample usually consists of dissolving the sample in an appropriate solvent. If the activity of the element of interest is large enough to be obtained by analysis of a composite decay curve, an aliquot of the solution containing the dissolved sample is transferred to a planchet, dried, and counted. However, if the activity to be counted is only a small fraction of the total activity or is mixed with another activity of about the same half-life, it is necessary to isolate the desired activity by a suitable chemical separation technique. A simple analytical group separation technique is sometimes adequate for this purpose. It is sometimes possible to determine the amounts of various components of a mixture directly, using a scintillation spectrometer.

It should be noted that inert carrier may be added after the irradiation to the solution of the irradiated sample before a general chemical separation is carried out. Hold-back carriers are also added for the other undesirable elements which have been activated. A similar chemical process is used for the irradiated reference sample. The comparison of the activities of the separated constituent of the sample with the reference one can get the amount of the sought element present in the original sample.

13.8.3 Advantages/Disadvantages of This Technique

1. An advantage of this method is that the sensitivity for different elements can be increased or decreased by a suitable choice of irradiation time.
2. It is highly selective because the choice of the carriers depends upon the element of the interest and has no effect on any other impurities present in the sample. However, the efficiency of the process and the purity of the element separated should always be checked by radiochemical methods.
3. Errors in the determination may be somewhat larger with species of low natural abundance and where poor counting statistics are obtained. The counting of low energy particulate radiation is sometimes a problem. There are also cases where more than one elements in the sample give rise to the same species on irradiation; for example, Aluminum-27 [by the reaction ^{27}Al (n, α) ^{24}Na] and Sodium-23 [by the reaction ^{23}Na (n, γ) ^{24}Na] both give Sodium-24 when irradiated in a reactor, which makes the determination of traces of sodium in aluminum impracticable by this method.
4. In the activation analysis, the neutron flux should be as high as possible, otherwise the activity produced due to a micro-component may be very low and not easy to detect and measure. This limits the use of ordinary laboratory neutron source, which usually gives a neutron flux of up to only about 10^5 neutron cm^{-2} sec^{-1}. On the other hand, neutron generators based on the following reactions are useful and give a neutron flux of about 10^9 neutron cm^{-2} sec^{-1}.

$$^3H_1 + {}^2H_1 \longrightarrow {}^4He_2 + {}^1n_0 + 14\,\text{MeV}$$

Hence fast as well as slow neutron activation analysis can be carried out. Since neutron generator can be installed in the laboratory, non-destructive experiments can be carried out with great precision. However, an adequate shielding of the generator is essential. In addition, the tritium target material has a short life of about 12 h and therefore, requires regular charging quite often. Both of these factors make the neutron generator expensive. Nevertheless, for shorter irradiation work neutron generators can be used.

Summary

In this chapter, we studied the necessities of the carrier of different types for performing any radiochemical separation techniques. Various types of most common radiochemical separation techniques are discussed. One of the common techniques used for the identification of metals present in any unknown sample is by activation analysis. The principle and its application has been discussed in the chapter.

Chapter 14
Hot Atom-Nuclear Reaction

14.1 Introduction

There are many good books dealing with Nuclear chemistry but very few deal with nuclear reactions which can be of interest to chemists, especially the effect of nuclear transformation on the chemical nature of the product formed during a nuclear reaction. In this chapter, we shall mainly be discussing this type of nuclear reaction. The study of the chemistry of atoms produced by nuclear reactions is often called "**Hot atom**" chemistry, and includes, for example, the effect of nuclear collisions on the chemical state of the target element. It is well established that when a neutron interacts with a target nucleus, some nuclear adjustment takes place inside the nucleus of the target and finally one or cascades of γ-rays of various energies are emitted. This emission causes the recoil to the target nucleus. The recoil energy is of the order of 100 eV which is much higher than the chemical bond energy (1–5 eV). Thus, the momentum imparted to a nucleus in a nuclear reaction with a charged particle or a fast neutron or a slow neutron is invariably sufficient to result in the rupture of any chemical bonds holding the atom in a molecule. A similar type of bond rupture has also been observed in chemical compounds containing radioactive atoms which decay either by α, β or by isomeric transition or by electron capture. It is, therefore, interesting to examine the consequence of such nuclear reactions on the chemical nature of products formed in a nuclear reaction.

14.2 Szilard-Chalmers Reactions

Fermi found that when a covalently bonded element (atomic number up to 30) is irradiated in a Ra/Be neutron source, a radioactive nuclide is obtained, which has different chemical properties from those of the parent and can easily be separated from the parent. Later, in 1935, Szilard in collaboration with Chalmers extended this work to other higher atomic number elements. They stated that, in general, the radioactive nuclide thus formed is an isotope of the target element and is in a free state, which can be separated easily. Ethyl iodide was the first compound studied by them. From the irradiated ethyl iodide, Iodine-128 was separated in the aqueous phase with a high specific activity by using chloroform and sodium thiosulfate solution as immiscible solvent (i.e., by solvent extraction technique). This reaction is named after the discoverers, the **Szilard-Chalmers reaction**.

The mechanism of this reaction can be explained by considering the laws of conservation of momentum. When a nucleus undergoes an (n, γ) reaction, the γ-rays (usually referred to as the γ-rays of capture) are emitted within the time of the order of 10 picosecond of neutron capture. The energy of the γ-rays equals the binding energy of a neutron within the nucleus and is usually of the order of 7–8 MeV. To conserve momentum, the nucleus must recoil with an energy which can be shown by a simple calculation to be, in general, greater than the binding energy of atoms in molecules.

The energy of γ-photon (E_γ in MeV) is given by hv, where h is a Planck's constant and v is the frequency of the photon. Momentum of the recoiling nucleus (p) is equal to the momentum of the photon and may be given as

$$p = mv = \frac{hv}{c} = \frac{E_\gamma}{c} \tag{14.1}$$

The energy of recoil E_R, is therefore

$$E_R = \frac{1}{2}(mv^2) = \frac{p^2}{2m} = \frac{E_\gamma^2 \times 536}{M} \text{ eV} \tag{14.2}$$

where m, c, M and 536 are mass of recoiling nucleus, velocity of light, atomic mass unit of recoiling atom and a conversion factor of units which also takes into account the velocity of light, respectively.

From this equation, the recoil energy E_R can be calculated, provided the energy of γ-photon is known. For example, the energy of the emitted γ-photon from bromine atom being 5.1 MeV, the recoil energy E_R from the Eq. (14.2) comes to 178 eV. A comparison of this recoil energy with the chemical bond energy (which is about 1–5 eV) shows that the separation of the radioactive nuclide thus formed is likely to be complete. However, in practice, usually only about 50–60% of the total activity is separable and the rest are found to be present in their original chemical form. The amount of activity, which remained with the parent chemical form, is also called the **retention activity**.

Scientists have been enquiring as to why only a very few percentage of total activity was available in the form, different from the parent compounds. Because one would expect to get 100% bond rupture in the nuclear recoil process. There have been several proposed mechanisms to explain the process by which the recoil atoms recombine with the target molecules. According to Libby, when the halogen atom of an alkyl-halide captures a neutron, it recoils with an energy considerably in excess of the chemical bond energy. It loses energy, largely by elastic collision (no transfer of energy) most effectively when it hits an atom of similar mass. If sufficient energy is lost from the radioactive atom during collision, the struck molecule gets disrupted and if the radioactive atom has insufficient energy after the collision to escape from the surrounding molecules (the so-called **molecular cage**), it is held in the vicinity of the molecular residue with a high probability of combination (i.e., retention of radioactive nuclide results). However, if the energy of the radioactive nuclide after collision is sufficiently large to escape from the cage, but not large enough to produce further dissociation, it will not be able to combine after another collision and it will emerge as a free halogen atom unless the side reaction occurs. Free halogen atom will then combine to form halogen molecules.

If an alkyl-halide is irradiated in the gas phase, a very low retention value is obtained supports the above statement, because the energy required to escape from the surrounding molecules in the gas at a site of collision is much less. Even by diluting the material with the solvent, the retention can be decreased, since the molecules are dispersed further away from each other.

In general, the Szilard-Chalmers reactions can satisfactorily be observed provided:

1. The recoil energy is greater than the chemical bond energy,
2. The radioactive atoms combine with the parent molecules inefficiently i.e., the retention value is low.
3. Chemical exchange does not take place between the stable and the radioactive species. If there is an exchange, the retention of activity will be appreciable. For example, exchange of stable bromine with the radioactive bromine in aromatic nitro halogen derivatives may result in a loss in the recovery of free radioactive halogen molecules produced in this reaction.
4. There is a suitable method for the separation of the radioactive nuclide existing in a different chemical state to the parent molecules.

14.2.1 Szilard-Chalmers Reaction with Organic Substances

A considerable amount of work has been done on the Szilard-Chalmers reaction with organic molecules, perhaps because the radioactive atom can be separated easily from the parent one. An interesting effect of this reaction is the production of labeled compounds with a high specific activity, such as bromobenzene, dichlorobenzene etc. This is very important because the production of high specific activity of a long-lived nuclide is not possible by ordinary irradiation methods. For example, 10% of the total

activity has been recovered in the form of chlorobenzene by irradiating a mixture
of carbon tetrachloride and benzene. It is also possible to obtain a different labelled
compound by irradiating a single species. For example, labelled dichlorobenzene can
be extracted from irradiated chlorobenzene. In such separations, compound carriers
are usually used to isolate the products or even to find out the various products
formed.

The production of such labelled compounds by an (n, λ) reaction can satisfactorily
be explained by considering the Libby theory of retention. In the previous discussion,
it was assumed that the breakage of a chemical bond takes place only between the
halogen atom and the immediately adjoining carbon atom. But this is not the only
possibility; the strength of the bond between, C–C or C–H is also of the order of
5–8 eV, and these bonds have a probability of being ruptured. If breakage of a C–C
bond takes place, the production of a new chemical species containing radioactive
halogen and fewer carbon atoms will result.

14.2.2 Szilard-Chalmers Reaction with
Inorganic Substances

Though the Szilard-Chalmers reaction with inorganic substances have been studied in detail, the interpretation of the results have been mainly speculative. This
is because, the chemical separation of the constituents from the solid substances
requires dissolution of the material after nuclear reaction. During the process of dissolution, many side reactions are initiated resulting in different species. Moreover, it
has not been established with certainty about all possible reactions occurring during
the dissolution. As a result of this uncertainty, the results with most of the inorganic
solid materials have always been difficult to reproduce. Nevertheless, work has been
done with phosphate, permanganate materials etc. For example, when a phosphate
solution is irradiated in a neutron flux, labelled phosphite is produced. The retention
of activity is about 50% irrespective of the nature of the target materials (i.e., whether
solid or either acidic or alkaline solution is irradiated). Because it is believed that the
initial disruption of phosphate may occur by either of the two almost equally probable routes, one leading back to phosphate and the other giving the lower valence
state (phosphite).

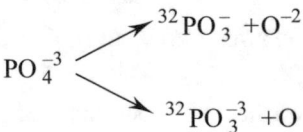

The separation of such products may be carried out by precipitating the expected
form by the addition of the appropriate species as carrier; there should be no chemical
exchange between the two chemical forms or any reaction with water during the
dissolution. The latter assumptions are very difficult to achieve.

14.3 Application of Szilard-Chalmers Reactions

- One of the most important applications of this reaction is the production of labelled inorganic or organic compounds at a high specific activity, as has been previously mentioned.
- Another application is to estimate an internal conversion coefficient (i.e., the ratio of the number of nuclides decaying by internal conversion to the total number of γ decaying). For example, during the irradiation of a bromo-compound in the neutron source, there is the possibility of two types of reactions:

$$R\text{-}Br \begin{cases} {}^{80m}Br\ (T_{0.5} = 44\ hr) + R\ (\sigma\ \text{for the reaction 2.9b}) \\ \\ {}^{80m}Br\ (T_{0.5} = 18\ min) + R\ (\sigma\ \text{for the reaction 8.5b}) \end{cases}$$

Both Bromine-80m and Bromine-80g will have some retention in the organic phase. If a separation of inorganic bromine is carried out initially and then again after an hour, the activity in the aqueous phase in the second extraction must be due to Bromine-80g, as a result of the following reaction:

$$R\text{-}Br^{80m} \xrightarrow{I.T.} R\text{-}Br^{80g} + h\upsilon$$

From the activity thus separated, it is possible to estimate the internal conversion coefficient. The isomeric transition does not provide enough recoil momentum to break the C–Br bond because of the low energy of γ-ray emission (0.037 MeV). However, Bromine-80m, also decays in internal conversion resulting in the emission of an electron from the K- or L-shell of the product atom i.e., from Bromine-80g. This may cause a valence electron to fall into a vacancy in either the K-shell or L-shell, leaving the atom positively charged. This would provide sufficient excitation to rupture the chemical bonds (i.e., the chemical bond between C–Br).

- Another use of this reaction is to verify a parent-daughter relationship when isomeric pairs are involved. The most conclusive way to prove that radioactive nuclide X is the parent of the second one (Y), is by means of the chemical separation of X from Y. If Y appears in X, then Y must be have come from X. But if, either of them is an isomer, then it is not possible to distinguish between the parent and the daughter. However, the Szilard-Chalmers reaction can sometimes give satisfactory proof. For example, in the previous example of bromine, activity extracted after one hour is due to the daughter form, and determination of half life of this form would confirm that nuclide decaying with 4.4 hrs half life is the parent of the isomeric pair i.e., ${}^{82m}Br$.

Summary

In this chapter, we learnt about a special type of nuclear reaction name as Szilard-Chalmers reaction. In this type of reaction, it is observed that after any type of nuclear reaction, the product formed normally emits some electromagnetic radiations like γ-rays. Emission of such radiations causes the product to experience a recoil of energy which is much greater than the chemical bond energy. Due to this recoil energy, many different types of products are formed. Theory of such a process is discussed. The advantage of this type of recoil reaction is that, one can produce radioactive isotopes with a high specific activity in time much shorter than one would need if it were to be formed by direct nuclear reaction. This aspect has also been discussed in this chapter.

Problem

In order to appreciate the various aspects of the topics discussed in this book, some useful questions are given here. It is hoped that while attempting these questions, some of the intricacies of the subject will get revealed in a better manner.

1. Explain the significance of the binding energy and the packing fraction in understanding the stability of a radioactive isotope.
2. For $^{238}U_{90}$, the binding energy per nucleon is 7.576 MeV. Calculate the mass of this isotope. Mass of hydrogen is 1.008145 amu and that of neutron is 1.008986 amu, where 1 amu = 931 MeV.
3. The following atomic masses are given below:

$$^{20}F_9 = 20.006341 \text{ amu}$$

$$^{20}Ne_{10} = 19.998765 \text{ amu}$$

$$^{20}Ne_{11} = 20.015236 \text{ amu}$$

   Based on the mass consideration, suggest which of these nuclides will decay by β-emission, γ-emission, and by orbital electron capture? Calculate the decay energy in MeV.
4. Explain the type of decay $^{32}P_{15}$ undergoes. Discuss the thermodynamic reasons for such type of decay. Masses of a few isotopes are given for you to substantiate your discussions. $^{32}P_{15} = 31.973908$ amu, $^{32}Si_{16} = 31.974020$ amu and $^{32}Si_{16} = 31.972074$ amu.
5. Distinguish among β^-, β^+, and k-capture decay processes. Explain the conditions under which radioactive isotopes can decay by these processes.
6. A certain radioactive substance (assume that it does not have any radioactive parent) has a half-life of 8.0 days. What fraction of the initial amount will be left after (i) 16 days and (ii) 39 days?
7. Define "Mass defect" and "Packing fraction". From the masses given below, calculate the binding energy in MeV for (i) an additional neutron to $^{239}Pu_{94}$ and (ii) an additional proton to $^{52}Mn_{25}$.

M. Sharon and M. Sharon, *Nuclear Chemistry*,
https://doi.org/10.1007/978-3-030-62018-9

Given:

$$^{52}Mn_{25} = 51.96202 \text{ a.m.u.}$$
$$^{239}Pu_{94} = 239.1265 \text{ a.m.u.}$$
$$^{240}Pu_{94} = 240.1296 \text{ a.m.u.}$$
$$\text{Proton} = 1.00814 \text{ a.m.u.}$$
$$\text{Neutron} = 1.00898 \text{ a.m.u.}$$

8. Explain the term "half-life" and "average life" of a radioactive isotope. How are they related to each other? Given that 1 g of $^{226}Ra_{88}$ emits 3.66×10^{10} α-particles per second and the half-life of radium is $1620y$, calculate the Avogadro number.

9. Discuss a suitable method for the determination of half-life of a radioactive isotope. Calculate the weight "w" in g needed to get 1.00 mCi of activity of ^{14}C. The half-life of the isotope is $5720y$.

10. Given that each atom of ^{238}U that decays gives ultimately one atom of ^{206}Pb, find the age of the mineral in which 1.33×10^{-2} g of ^{206}Pb is associated with each gram of ^{238}U. The half-life period of ^{238}U is 4.5×10^9.

11. Discuss the law of radioactive decay. Explain the relationship among the decay constant, the half-life, and the mean life of a radioactive isotope. Calculate the activity of 1.0×10^{-6} g of ^{24}Na in curies, given that $T_{0.5}$ is 14.9 h, atomic weight of ^{24}Na is 24.00, 1 Ci is 3.7×10^{10} dps, and Avogadro number is 6.023×10^{23}.

12. Calculate the amount of ^{210}Po (in g) that can be obtained from 1 g of ^{226}Ra, assuming the sample of radium is more than $100y$ old. The half-life of ^{226}Ra is $1620y$ and half-life of ^{210}Po is $138d$.

13. A radioactive element X decays to another radioactive element Y. If λx and λy are their decay constants, write the expression for the activity of Y at time t. Deduce the conditions for secular and transient equilibria. Explain their significance in measuring activities of radioactive isotopes undergoing such type of equilibrium.

14. Given that the ^{14}C content in the plant normally gives a specific activity of 15.3 cpm per gram, find the age of the wood of the plant which gives the specific activity of 5.3 cpm per gram. $T_{0.5}$ of ^{14}C is $5720y$.

15. The half-life of ^{221}Fr is 4.8 min. Starting with one mg of isotope, how much would remain after 30 min?

16. Explain the following:

    (a) The average life is greater than half-life by a factor.
    (b) The nuclei $A \geq 140$ decays by α-emission rather than proton emission.

17. β-radiation from a source is to be measured with a G.M. counter. The maximum uncertainty permitted is $\pm 1.0\%$. The counter records at the end of successive 5 min periods for sample and background are as follows:

Time/min	0	5	10	20	25	30
Background counts	0	127	249	502	672	793
Sample count	0	2155	4297	8602	10749	12907

(a) What is the minimum time over which the counting must be done to give the required statistical precision?

(b) How long should the background count be counted?

(c) What is the activity in the unit of counts per minute the precision limits?

18. (a) Explain the principle of scintillation counter and discuss its application in the determination of the energy of γ-rays of an isotope in solid form.

(b) Given a proportional counter and liquid scintillation counter, explain with reasons which of them is preferable for counting the activity of a precipitate of $BaSO_4$ labeled with ^{35}S (E_{max} for the β-particle is 0.167 MeV).

19. Discuss the principle and design of a gas flow proportional counter. Explain the relative merits of counting ^{14}C (β-particle energy 0.145 MeV) by the following counters:

(a) Thin mica window halogen quenched Geiger–Müller counter.

(b) Gas flow proportional counter.

(c) Plastic phosphor scintillation counter.

20. (a) Discuss the principle and operation of a gas flow proportional counter and compare its features with those of an end-window G.M. counter.

(b) What is a pulse height analyzer? Explain how it is useful in measuring the energy of β-rays of ^{60}Co.

21. (a) Explain what you understand by "resolution time" of a counter and its effect on the count rate.

(b) Give an account of characteristic features of an ionization detector to be used for the detection of α-particles and β-particles.

22. Considering the decay scheme of ^{22}Na, which type of counter will you select to count a microcurie of this isotope? In what form would you prefer to count, i.e., in solid or liquid, and why? Discuss the relative reasons for your selection of the counter. Give the theory of the counter selected by you for this purpose.

23. Argon gas enclosed in a cylindrical vessel is exposed to a particulate radiation. Discuss the current–voltage characteristics of this gas and explain its application in designing various ionization counters for the detection and measurement of radioactivity.

24. "Proportional counter can distinguish pulses originated from α-particles and β-particles and not a G.M. counter". Explain.

25. Discuss the principle and operation of a gas flow proportional counter and compare its characteristics with that of a sealed gas proportional counter.

26. "In a proportional counter, a pulse height analyzer can be used to determine the energy of α-particles as well as E_{max} for β-particles". Explain.

27. What is the dead time of a counter and from where does it originate? Why does the G.M. counter have a dead time of the order of 200 ms, whereas the proportional counter has 1–10 ms? Explain.

28. Explain the working principle of a Geiger–Müller counter. G.M. (end-window) counter can be used to detect β-particles and γ-rays and not α-particles. Why? Explain.

29. What is a plateau of an ionization counter? What information do you derive out of the current–voltage characteristics of an ionization counter?

30. Semiconductor detectors are sensitive to detect α-particles efficiently and the level of background activity recorded by such counter is very low. Why? Explain.

31. A Geiger–Müller counter can be made of any shape and size, whereas there is some restriction to the design of proportional counter. Is it a correct statement? Discuss.

32. What are the advantages of a halogen quenched Geiger–Müller counter over the organic quenched G.M. counter? Explain the mechanism of quenching spurious pulses with this type of counter.

33. How will you determine the dead time of a liquid G.M. counter? Would it differ from one counter to another? If so, why? Explain. What is the best method for correcting the dead time of a G.M. counter? Explain.

34. Explain the principle of scintillation counter and explain with the block diagram, a scintillation counter for determining the γ-rays spectrum of a radioactive source.

35. Why is NaI(Tl) used for counting high energetic-β-particles and γ-rays and not for α-particles, whereas organic scintillators are used for β-particles and α-particles but not *vice versa*?

36. How will you find out the best operating conditions for obtaining an energy spectrum of γ-rays by the scintillation counter? How does this differ from the operating conditions for counting β-particles?

37. The identification of a γ-emitting source and its characterization can be done accurately with the scintillation counter and not with a proportional counter. Discuss.

38. Explain the use of a pulse height analyzer in the scintillation counter to identify a γ-source.

39. Discuss the application of standard deviation calculation in radioactive measurements and explain why a mean count rate of 100 obtained from a population of count rates of range 80–120 will have 30% error? What procedure would you adopt to decrease this percentage of the error to less than 1%, and what are the factors you would need to consider for achieving this accuracy of counting?

40. Which counter would you use to detect the isotope after considering the decay scheme of a Cesium-137 isotope? Explain the reasons for your selection.

41. (a) Explain the principle of a scintillation counter and discuss its application in the determination of the energy of γ-rays.

(b) 1.25 g of calcium iodide was irradiated in a nuclear reactor for two hours. Calculate the activity of the Iodine-128 isotope produced. What would be the specific activity of this isotope?

Given: cross-section for Iodine-128 is 6.4 barns, thermal neutron flux is 3.0×10^{11} neutron $cm^{-2}s^{-1}$, and half-life of Iodine-128 is 26.5 min.

42. Explain the principle of a scintillation counter. Describe how the pulse height analyzer is used together with the scintillation counter to measure the energy of γ-rays.

43. To a crude mixture of organic compounds containing some benzoic acid and bezoate 40.0 mg of benzoic acid labeled with ^{14}C (activity 2000.0 cpm) was added. After equilibrium, the mixture was extracted with immiscible solvents. The extracted solid, following the removal of the solvent, was purified by the recrystallization of the benzoic acid to a constant melting point. The purified material weighed 60.0 mg and gave a count rate of 500 cpm. Compute the weight of benzoic acid in the mixture.

44. How does an ionization counter differ from a scintillation counter? Discuss in five lines.

45. What are the basic differences among a G.M. counter, a proportional counter, and an ionization counter?

46. Which counter will you select to count the activity of a liquid sample giving 100% of energy 0.55 MeV as well as 100% particles of E_{max} 0.55 MeV and why?

47. How can a mixture of radioactive sample emitting an α-particle and a β-particle of the same energy (0.98 MeV) be counted by a proportional counter without performing any chemical separation while it is not possible with an end-window mica G.M. counter? Explain and discuss.

48. Why does a G.M. counter need no amplification of signals as needed in a proportional counter? Why is a pulse height analyzer not used in a G.M. counter to differentiate high energetic β-particles from γ-rays? Explain.

49. Why does one find out a plateau in an ionization-type counter and why does a proportional counter give two plateaus, one for α-particles and another for β-particles?

50. What information do you get from the plateau of a G.M. counter? A G.M. counter has a dead time of 450 ms. A Cesium-137 isotope was counted for 10 min and it gave an activity of 10,000 counts. Calculate the % loss of activity with this G.M. counter.

51. What information regarding nuclear energy levels is obtained by the measurement of the energies of β-particles and γ-radiation in a β-decay? How does it help in understanding the decay scheme of a radioactive isotope?

52. A ^{27}Mg isotope undergoes β-decay and in about 70% of the disintegration, the β-particles have E_{max} of 1.78 MeV while in about 30% of the disintegration the E_{max} is 1.59 MeV. γ-rays of energies 0.834 MeV, 1.015 MeV, and 0.181 MeV are also observed to be emitted during the disintegration. Construct a decay scheme consistent with these observations.

53. What are the mechanisms by which energy is transferred to the material irradiated by electromagnetic radiations of different energies? Distinguish among the linear, mass, and atomic absorption coefficient of an electromagnetic radiation in matter.

54. What is Compton scattering? How does the cross-section for the process depend on the nature of the material and the energy of the photons?

55. What aspects of the interaction of electromagnetic radiation with matter have led to the development of the scintillation counter. Discuss.

56. (a) Distinguish between the modes of interaction of α-particles and γ-rays with matter. How do these interactions help in detecting these radiations?

    (b) γ-radiations of 1 MeV undergo Compton scattering in a material. Calculate the maximum energy loss in the radiation in a single interaction if the Compton wavelength is 0.024°A. What would be the condition when there is no loss of energy by Compton scattering?

57. (a) Describe the processes by which γ-radiation loses its energy while passing through the matter.

    (b) Given the half-thickness value for water for γ-rays emitted by ^{60}Co as 11 cm, calculate the mass absorption coefficient of water. Calculate the thickness of water that would be required to reduce the intensity of a 1000 Ci source of ^{60}Co by a factor of 10. (Density of water is 0.998 g cm^{-2}.)

58. Explain the principle of a proportional counter and compare its merit with that of the liquid scintillation counter.

59. What are the different types of ionization counters? Explain the applications and limitations of these detectors.

60. (a) Describe the design and operation of a scintillation counter. Explain how is it used for measuring β-particles and γ-radiations.

    (b) A counting system has a resolving time of 150 ms. The observed count rate is 10,109 cpm. Calculate the true count rate.

61. Find out the decay scheme of ^{90}Sr. A sample of freshly prepared ^{90}Sr is given to you. Would you prefer to count the sample after 10–12 days of separation or immediately after the separation to get the best efficiency in counting? Give reasons for your answer.

62. Find out the decay scheme of ^{111}Ag. Which counter would you use to count this isotope to get a good efficiency of counting? Give the theory of the counter selected by you.

63. What modification (*i.e.*, coincidence or anti-coincidence) would you like to do with your selected counter to increase the efficiency of the counting and why?

64. Describe the design and the operation of a scintillation spectrometer. Explain how it is used for measuring α-particles, β-particles, and γ-rays.

65. A counting system has a resolving time of 500 ms. The observed counting rate is 1250 cpm. Calculate the true counting rate and coincidence error.

66. What are the essential differences (*i*) between an ionization chamber and a counter and (*ii*) between a proportional counter and a Geiger–Müller counter?

67. Explain the principles of the scintillation counter and its use for measuring α-particles, β-particles of energy 1.58 MeV, and γ-rays of 1.2 MeV.

68. Describe the recent developments in the detection and measurement of radia-
    tions. What is the smallest amount of ^{198}Au expressed in microcuries that you
    can measure with $\pm 10\%$ standard deviation? Assume that you are provided with
    a well-type scintillation counter which gives an efficiency of 25.6%. Let each
    counting be done for 10 minutes. The background count was 653 obtained in 10
    min of counting.

69. Find out the decay scheme for ^{64}Cu:
    The sample is present in a solid form. How will you prepare the source for
    counting this isotope and which counter would you select for counting purposes?
    Explain.

70. Find out the decay scheme of ^{90}Y. The mass of ^{90}Zr is 89.9329 amu. What is the
    mass of ^{90}Y?

71. Discuss the principle and applications of a coincidence and anti-coincidence
    type of counters. Considering the decay schemes, explain which counter you
    would select to count ^{63}Ni and ^{38}Cl isotopes separately.

72. Why is a photomultiplier tube used in a scintillation counter? Can you replace
    the photomultiplier tube with something to measure α-particles giving activity
    1.05 cpm, but without using any other method of counting, i.e., proportional,
    G.M., or solid-state counters?

73. Considering the decay scheme of ^{111}Ag, which counter will you select for count-
    ing this isotope in liquid form? Discuss by giving reasons for your selection of
    the counter.

74. A sample of freshly prepared ^{90}Sr is provided to you. How will you count this
    isotope and in which form would you prefer to measure the activity? Would you
    prefer to count the sample immediately or after one week? Give reasons for your
    choice. Explain the decay scheme of this isotope.

75. Find out the decay scheme of ^{42}K:
    Which type of radiations emitted by this isotope would you prefer to use for
    counting this isotope and why?

76. Suggest which counter you would prefer to count the following radioactive iso-
    topes. Give reasons for your answer.

    (a) ^{24}Na in solid form (E_{max}) for β-particles 1.39 MeV and 1.36 MeV, $T_{0.5}$ 15 h.
    (b) Uranium mixed with ^{24}Na in solid form.
    (c) Tritium in the form of aqueous solution (E_{max}) for β-particles 0.018 MeV
        $T_{0.5}$ 12.35y. What considerations would you make if this isotope is to be
        counted in solid form?

77. Why do α-decay and high energy nuclear reactions leave the parent atom in
    an excited state, whereas β-decay, EC-decay, or IT-type decay do not? Would
    you agree that in all the above cases, the daughter molecule will have a form
    chemically different from the parent and if so, why?

78. Explain the various considerations you would make for counting an α-emitting
    source and an energetic β-emitter source.

79. What do you understand by self-absorption? What will be its effect in counting
    the radiation of a source emitting low energy β-particles?

80. What is the difference between electro-deposition and electro-spraying techniques? In the electro-spraying method, what will happen if your substance is hygroscopic? Explain the effect of the source being non-uniform especially for counting low energetic β-particles.

81. How does the geometry of the counter affect spurious pulses? And what are the factors on which they depend?

82. What are the applications of a $4\pi\gamma$-counter? What is meant by coincidence counting? Describe the design of a counter to measure an activity of 2–6 cpm. Describe the arrangements needed to introduce into a counter to reduce the spurious pulses down to almost zero.

83. Explain the principle of a semiconductor detector used for measuring the activity of α-particles. Explain its advantages over other counters. Why is the background activity in such counter very low even without using any lead shielding? Explain.

References

Some useful references are given here. But these should not be taken as an exhaustive list of all the concerned references. Much information and examples given in the text are the results of the personal experience of the author.

Atac M. and Taylor W. E., *High resolution cryogenic spark chambers*, Nucl. Inst. & Methods, 118, 413–417 (1974).

Bambynek W., *On selected problems in the field of proportional counters*, Nucl. Inst. & Methods, 112, 103–110 (1973).

Berom B. L., Crawford J. F., Ford R. L., Hofstaddtler R., Huges E. B., Kose R., Le-Coultre, Martin T. W., Nei L. H. O., Rand R. E., Schilling R. F. and Wedemeyer R., *Recent applications of NaI(Tl) total absorption counting*, HEPL Report No. 712 June 1–2 (1973).

Bokowski D. L., *Liquid scintillation counting for plutonium in environmental samples*, Am. Industrial Hygiene Association J., June 333–344 (1974).

Camion P. J., *Spurious pulses in proportional counters - A review*, Nucl. Inst. & Methods, 112, 75–81 (1973).

Choppin GRC *et al. Separation of transuranic elements by ion-exchange method*, J. Am. Chem. Soc., 76, 6229 (1954).

Cavallo L.M., Coursey B. M., Garinkel S. B., Hutchinson J. M. R. and Mann W. B., *Needs for radioactivity standards and measurements in different fields*, Nucl. Inst. & Methods, 112, 5–18 (1973).

Cook & Duncon, *Modern Radiochemical Practice*, The Claredon Press Oxford 1958.

Danish Taylor *Measurements of radioisotopes*, Methuen and Co. Ltd., London 1959.

Eric Schram, *Organic Scintillation detectors Elsevier*, Pub. Comp. 1963.

Friendlander and Kennedy, *Nuclear and Radiochemistry*, Capmann Hall Ltd., 1955.

Glenn F. Knoll, *Radiation detection and measurement*, John Wiley & Sons, Singapore, 1989.

Gregory R. Choppin, *Experimental Nuclear Chemistry*, Prentic Hall Inc., 1961.

Gyonyor B. A., *A low background β-counting arrangement*, Acta Physica Academiae Scientrarum Hungaricae Tomus, 35(1–4), 131–139 (1974).

Hassib G. M. and Medveczky L., *New characteristics for cellulose nitrate film as a neutron track detection*, Atomic Kozlemeyek, 16/4, 315–324 (1974).

Hiromichi Nakahara, Minoura Yanakyra, Yoshihiro Nakagome and Tetsuo Hashimoto, *Thermal neutron induced fission of Radium-226*, J. Inorg. Nucl. Chem. 38, 203–204 (1974).

Hutchinson J. M. R.; Mann W. B. and Perkina R. W., *Low-level radioactivity measurements*, Nucl. Inst. & Methods, 112, 229–238 (1973).

Iyer R. H., *Solid-state track detectors* J. Chemical Education, 49(11), 742–745 (1972).

© The Author(s), 2021

M. Sharon and M. Sharon, *Nuclear Chemistry*,

https://doi.org/10.1007/978-3-030-62018-9

Legrand J., *Calibration of γ-rays*, Nucl. Inst. & Methods, 112, 239–248 (1973).

Leon Grigorescu, *Accuracy of coincidence measurements*, Nucl. Inst. & Methods, 112, 151–155 (1973).

Martin L. Nusynuwitz and Anthony R. Benedetto *Simplified determination of radioactive decay factors*, J. Nucl. Medicine Technology, 2(3), 99–101 (1974).

Nesmeyyanove A. N., *A guide to practical radiochemistry*, Mir Publ. Moscow 1984.

Ouseph P. J. and Schwartz M., *LXXV Nuclear Radiation Detectors*, J. Chemical Education, 51(3), 139–209 (1974).

Pahor J. and Hribar M., *Multiwire Counting*, Nucl. Inst. & Methods, 112, 279–283 (1973).

Ralph T. Overmann, *Basic concepts of Nuclear Chemistry*, Reinhold Publ. Corp. 1963.

Rapkin E. R. *Development of the modern liquid scintillation counter-the current status of liquid scintillation counting*, Ed. Edwin D. Bransome Jr. M. D., Grum & Stratton Inc. N.Y. pp. 450–68 (1970).

Sidhu N. P. S., *Stabilization of scintillation γ-ray spectrometer*, BARC/1-218, 1–5 (1972).

Somogy G. *Influence of thermal effect on the track registration characteristics of plastics Radiation Effect*, 16, 245–251 (1972).

Spernol A., *Some general aspects of radioactivity measurement methods*, Nucl. Inst. & Methods, 112, 23–31 (1973).

Steyn J. *Tracer method for pure β-emitter measurement*, Nucl. Inst. & Methods, 112, 157–163 (1973).

Tait W. H., *The spark chamber*, J. Inst. Nucl. Eng., 14(6), 171–179 (1973).

Wahl and Banner, *Radioactivity Applied to Chemistry*, John Willey and Sons Inc. 1958.

Weiss H. M., 4pg-*ionization chamber Measurements*, Nucl. Inst. & Methods, 112, 291–297 (1973).

White House and Putman, *Radioactive Isotope*, The Clardon Press, Oxford, 1958.

Wivan der Eijk, Oldenhof W. and Zehner W. *Preparation of thin sources a review*, Nucl. Inst. & Methods, 112, 343–351 (1973).

Wolfgang Richard, *Nuclear recoil as a means of fission product separation*, J. Inorg. Nucl. Chem. 180–183 (1956).

Index

© The Author(s), 2021
M. Sharon and M. Sharon, *Nuclear Chemistry*,
https://doi.org/10.1007/978-3-030-62018-9

233

Printed in the United States
by Baker & Taylor Publisher Services